Sustainable Forestry

Sustainable Forestry

Edited by
Eric Richardson

Larsen & Keller
www.larsen-keller.com

Sustainable Forestry
Edited by Eric Richardson
ISBN: 978-1-63549-271-2 (Hardback)

Larsen & Keller
Published by Larsen and Keller Education,
5 Penn Plaza,
19th Floor,
New York, NY 10001, USA

Cataloging-in-Publication Data

Sustainable forestry / edited by Eric Richardson.
p. cm.
Includes bibliographical references and index.
ISBN 978-1-63549-271-2
1. Sustainable forestry. 2. Forests and forestry. 3. Deforestation. 4. Forest management.
5. Forest conservation. I. Richardson, Eric.
SD387.S87 S87 2017
634.92--dc23

The publisher's policy is to use permanent paper from mills that operate a sustainable forestry policy. Furthermore, the publisher ensures that the text paper and cover boards used have met acceptable environmental accreditation standards.

Printed and bound in China.

For more information regarding Larsen and Keller Education and its products, please visit the publisher's website www.larsen-keller.com

Table of Contents

Preface

This book provides comprehensive insights into the field of sustainable forestry. It discusses in detail some of the essential functions and requirements for this subject. Sustainable forestry refers to the optimum use of forest resources. It incorporates areas like biodiversity conservation, protecting local livelihoods and forest ecosystems, reducing the negative effects of climate change, etc. This text attempts to understand the concerns regarding conservation that make up the discipline of sustainable forestry and how such concepts have practical applications. Such selected concepts that redefine this area have been presented in the text. It is meant for students who are looking for an elaborate reference text on sustainable forestry.

A short introduction to every chapter is written below to provide an overview of the content of the book:

Chapter 1 - The management and conservation of forests is known as forestry. The elements involved in it are biological, physical, social and political. It focuses on the provision of timber, wildlife habitat, recreating and quality management. This chapter will provide an integrated understanding of forestry; **Chapter 2** - Forests that receive more than 2000 mm of rainfall are known as tropical evergreen forests. The rainfall in rainforests varies from 250 to 450 centimeters. The other types of forests mentioned in this text are temperate coniferous forest, temperate deciduous forest, freshwater swamp forest and tropical and sub-tropical moist broadleaf forests. Forests can be best understood in confluence with the major types listed in the following section; **Chapter 3** - Forest management manages the administration, economy, the social aspects and technical aspects of forestry. The restoration of forestry is defined as ecoforestry. Ecoforestry takes a holistic approach in preserving forests and the organisms in it. The aspects elucidated in this chapter are of vital importance, and provide a better understanding of forest management; **Chapter 4** - Forests need to be surveyed and monitored and their growth and health needs. Some of the aspects of silviculture explained in the following chapter are tree planting, plant nursery, thinning, pruning and clearcutting. The topics discussed in the text are of great importance to broaden the existing knowledge on silviculture; **Chapter 5** - The elimination of a forest or of a number of trees is known as deforestation. Deforestation is mostly done for the purpose of urban use and the conversion of forestland to farms. Illegal logging, deforestation and climate change and forest dieback are some of the issues discussed within this chapter; **Chapter 6** - Taking measure for deforestation is very necessary, considering the issues deforestation can cause. Forest restoration and afforestation are the main ways of measurement. While forest restoration is the recovery of forest structures, afforestation is the installation of plants in an area that has previously had no trees. This section serves as a source in understanding the measures taken for deforestation; **Chapter 7** - The initiative taken for forest conservation is world forestry congress, billion-tree campaign and international day of forests. One of the largest assemblies of the world's forestry sector is the world forestry congress. The billion-tree campaign was introduced because of the alarming increase in global warming and climate change. The target of this challenge has been to plant one billion trees. This chapter discusses topics that are of great importance to broaden the existing knowledge on forest conservation.

I extend my sincere thanks to the publisher for considering me worthy of this task. Finally, I thank my family for being a source of support and help.

Editor

1

Understanding Forestry

The management and conservation of forests is known as forestry. The elements involved in it are biological, physical, social and political. It focuses on the provision of timber, wildlife habitat, recreating and quality management. This chapter will provide an integrated understanding of forestry.

Forestry

Forestry is the science and craft of creating, managing, using, conserving, and repairing forests and associated resources to meet desired goals, needs, and values for human and environment benefits. Forestry is practiced in plantations and natural stands. The science of forestry has elements that belong to the biological, physical, social, political and managerial sciences.

Forestry work in Austria

Modern forestry generally embraces a broad range of concerns, in what is known as multiple-use management, including the provision of timber, fuel wood, wildlife habitat, natural water quality management, recreation, landscape and community protection, employment, aesthetically appealing landscapes, biodiversity management, watershed management, erosion control, and preserving forests as 'sinks' for atmospheric carbon dioxide. A practitioner of forestry is known as a forester. Other terms are used a verderer and a silviculturalist being common ones. Silviculture is narrower than forestry, being concerned only with forest plants, but is often used synonymously with forestry.

Forest ecosystems have come to be seen as the most important component of the biosphere, and forestry has emerged as a vital applied science, craft, and technology.

Forestry is an important economic segment in various industrial countries. For example, in Germany, forests cover nearly a third of the land area, wood is the most important renewable resource, and forestry supports more than a million jobs and about billion in yearly turnover.

A deciduous beech forest in Slovenia

History

Background

The preindustrial age has been dubbed by Werner Sombart and others as the 'wooden age', as timber and firewood were the basic resources for energy, construction and housing. The development of modern forestry is closely connected with the rise of capitalism, economy as a science and varying notions of land use and property.

Roman Latifundiae, large agricultural estates, were quite successful in maintaining the large supply of wood that was necessary for the Roman Empire. Large deforestations came with respectively after the decline of the Romans. However already in the 5th century, monks in the then Byzantine Romagna on the Adriatic coast, were able to establish stone pine plantations to provide fuelwood and food. This was the beginning of the massive forest mentioned by Dante Alighieri in his 1308 poem Divine Comedy.

Similar sustainable formal forestry practices were developed by the Visigoths in the 7th century when, faced with the ever increasing shortage of wood, they instituted a code concerned with the preservation of oak and pine forests. The use and management of many forest resources has a long history in China as well, dating back to the Han Dynasty and taking place under the landowning gentry. A similar approach was used in Japan. It was also later written about by the Ming Dynasty Chinese scholar Xu Guangqi (1562–1633).

In Europe, land usage rights in medieval and early modern times allowed different users to access forests and pastures. Plant litter and resin extraction were important, as pitch (resin) was essential for the caulking of ships, falking and hunting rights, firewood and building, timber gathering in wood pastures, and for grazing animals in forests. The notion of "commons" (German "Allmende") refers to the underlying traditional legal term of common land. The idea of enclosed private property came about during modern times. However, most hunting rights were retained by members of the nobility which preserved the right of the nobility to access and use common land for recreation, like fox hunting.

Early Modern Forestry Development

Timber harvesting, as here in Finland, is a common component of forestry

Hans Carl von Carlowitz

Systematic management of forests for a sustainable yield of timber is said to have begun in the German states in the 14th century, e.g. in Nuremberg, and in 16th-century Japan. Typically, a forest was divided into specific sections and mapped; the harvest of timber was planned with an eye to regeneration. As timber rafting allowed for connecting large continental forests, as in south western Germany, via Main, Neckar, Danube and Rhine with the coastal cities and states, early modern forestry and remote trading were closely connected. Large firs in the black forest were called „Holländer“, as they were traded to the Dutch ship yards. Large timber rafts on the Rhine were 200 to 400m in length, 40m in width and consisted of several thousand logs. The crew consisted of 400 to 500 men, including shelter, bakeries, ovens and livestock stables. Timber rafting infrastructure allowed for large interconnected networks all over continental Europe and is still of importance in Finland.

Starting with the sixteenth century, enhanced world maritime trade, a boom in housing construction in Europe and the success and further Berggeschrey (rushes) of the mining industry increased timber consumption sharply. The notion of 'Nachhaltigkeit', sustainability in forestry, is closely connected to the work of Hans Carl von Carlowitz (1645–1714), a mining administrator in Saxony. His book *Sylvicultura oeconomica, oder haußwirthliche Nachricht und Naturmäßige Anweisung zur wilden Baum-Zucht* (1713) was the first comprehensive treatise about sustainable yield forestry. In the UK, and to an extend in continental Europe, the enclosure movement and the clearances favored strictly enclosed private property. The Agrarian reformers, early economic writers and scientists tried to get rid of the traditional commons. At the time, an alleged tragedy of the commons together with fears of a Holznot, an imminent wood shortage played a watershed role in the controversies about cooperative land use patterns.

The practice of establishing tree plantations in the British Isles was promoted by John Evelyn, though it had already acquired some popularity. Louis XIV's minister Jean-Baptiste Colbert's oak Forest of Tronçais, planted for the future use of the French Navy, matured as expected in the mid-19th century: "Colbert had thought of everything except the steamship," Fernand Braudel observed. In parallel, schools of forestry were established beginning in the late 18th century in Hesse, Russia, Austria-Hungary, Sweden, France and elsewhere in Europe.

Forest Conservation and Early Globalization

During the late 19th and early 20th centuries, forest preservation programs were established in British India, the United States, and Europe. Many foresters were either from continental Europe (like Sir Dietrich Brandis), or educated there (like Gifford Pinchot). Sir Dietrich Brandis is considered the father of tropical forestry, European concepts and practices had to be adapted in tropical and semi arid climate zones. The development of plantation forestry was one of the (controversial) answers to the specific challenges in the tropical colonies. The enactment and evolution of forest laws and binding regulations occurred in most Western nations in the 20th century in response to growing conservation concerns and the increasing technological capacity of logging companies. Tropical forestry is a separate branch of forestry which deals mainly with equatorial forests that yield woods such as teak and mahogany.

Mechanization

Forestry mechanization was always in close connection to metal working and the development of mechanical tools to cut and transport timber to its destination. Rafting belongs to the earliest means of transport. Steel saws came up in the 15th century. The 19th century widely increased the availability of steel for whipsaws and introduced Forest railways and railways in general for transport and as forestry customer. Further human induced changes, however, came since World War II, respectively in line with the '1950s-syndrome'. The first portable chainsaw was invented in 1918 in Canada, but large impact of mechanization in forestry started after World War II. Forestry Harvesters are among the most recent developments. Although drones, planes, laser scanning, satellites and robots also play a part in forestry.

Forestry Today

A modern sawmill

Today a strong body of research exists regarding the management of forest ecosystems and genetic improvement of tree species and varieties. Forestry also includes the development of better methods for the planting, protecting, thinning, controlled burning, felling, extracting, and processing of timber. One of the applications of modern forestry is reforestation, in which trees are planted and tended in a given area.

Trees provide numerous environmental, social and economic benefits for people. In many regions the forest industry is of major ecological, economic, and social importance. Third-party certification systems that provide independent verification of sound forest stewardship and sustainable forestry have become commonplace in many areas since the 1990s. These certification systems were developed as a response to criticism of some forestry practices, particularly deforestation in less developed regions along with concerns over resource management in the developed world. Some certification systems are criticized for primarily acting as marketing tools and lacking in their claimed independence.

In topographically severe forested terrain, proper forestry is important for the prevention or minimization of serious soil erosion or even landslides. In areas with a high potential for landslides, forests can stabilize soils and prevent property damage or loss, human injury, or loss of life.

Public perception of forest management has become controversial, with growing public concern over perceived mismanagement of the forest and increasing demands that forest land be managed for uses other than pure timber production, for example, indigenous rights, recreation, watershed management, and preservation of wilderness, waterways and wildlife habitat. Sharp disagreements over the role of forest fires, logging, motorized recreation and other issues drives debate while the public demand for wood products continues to increase.

Foresters

Foresters of UACh in the Valdivian forests of San Pablo de Tregua, Chile

Foresters work for the timber industry, government agencies, conservation groups, local authorities, urban parks boards, citizens' associations, and private landowners. The forestry profession includes a wide diversity of jobs, with educational requirements ranging from college bachelor's degrees to PhDs for highly specialized work. Industrial foresters plan forest regeneration starting with careful harvesting. Urban foresters manage trees in urban green spaces. Foresters work in tree nurseries growing seedlings for woodland creation or regeneration projects. Foresters improve tree genetics. Forest engineers develop new building systems. Professional foresters measure and model the growth of forests with tools like geographic information systems. Foresters may combat insect infestation, disease, forest and grassland wildfire, but increasingly allow these natural aspects of forest ecosystems to run their course when the likelihood of epidemics or risk of life or property are low. Increasingly, foresters participate in wildlife conservation planning and watershed protection. Foresters have been mainly concerned with timber management, especially reforestation, maintaining forests at prime conditions, and fire control.

Forestry Plans

Foresters develop and implement forest management plans relying on mapped resource inventories showing an area's topographical features as well as its distribution of trees (by species) and other plant cover. Plans also include landowner objectives, roads, culverts, proximity to human habitation, water features and hydrological conditions, and soils information. Forest management plans typically include recommended silvicultural treatments and a timetable for their implementation. Application of digital maps in Geographic Informations systems (GIS) that extracts and integrates different information about forest terrains, soil type and tree covers, etc. using, e.g. laser scanning, enhances forest management plans in modern systems.

Forest management plans include recommendations to achieve the landowner's objectives and desired future condition for the property subject to ecological, financial, logistical (e.g. access to resources), and other constraints. On some properties, plans focus on producing quality wood products for processing or sale. Hence, tree species, quantity, and form, all central to the value of harvested products quality and quantity, tend to be important components of silvicultural plans.

Good management plans include consideration of future conditions of the stand after any recommended harvests treatments, including future treatments (particularly in intermediate stand treatments), and plans for natural or artificial regeneration after final harvests.

The objectives of landowners and leaseholders influence plans for harvest and subsequent site treatment. In Britain, plans featuring "good forestry practice" must always consider the needs of other stakeholders such as nearby communities or rural residents living within or adjacent to woodland areas. Foresters consider tree felling and environmental legislation when developing plans. Plans instruct the sustainable harvesting and replacement of trees. They indicate whether road building or other forest engineering operations are required.

Agriculture and forest leaders are also trying to understand how the climate change legislation will affect what they do. The information gathered will provide the data that will determine the role of agriculture and forestry in a new climate change regulatory system.

Forestry as a Science

Over the past centuries, forestry was regarded as a separate science. With the rise of ecology and environmental science, there has been a reordering in the applied sciences. In line with this view, forestry is one of three primary land-use sciences. The other two are agriculture and agroforestry. Under these headings, the fundamentals behind the management of natural forests comes by way of natural ecology. Forests or tree plantations, those whose primary purpose is the extraction of forest products, are planned and managed utilizing a mix of ecological and agroecological principles.

Education

History of Forestry Education

The first dedicated forestry school was established by Georg Ludwig Hartig at Hungen in the Wetterau, Hesse, in 1787, though forestry had been taught earlier in central Europe, including at the University of Giessen, in Hesse-Darmstadt.

In Spain, the first forestry school was the Forest Engineering School of Madrid (Escuela Técnica Superior de Ingenieros de Montes), founded in 1844.

The first in North America, the Biltmore Forest School was established near Asheville, North Carolina, by Carl A. Schenck on September 1, 1898, on the grounds of George W. Vanderbilt's Biltmore Estate. Another early school was the New York State College of Forestry, established at Cornell University just a few weeks later, in September 1898. Early 19th century North American foresters went to Germany to study forestry. Some early German foresters also emigrated to North America.

In South America the first forestry school was established in Brazil, in Viçosa, Minas Gerais, in 1962, and moved the next year to become a faculty at the Federal University of Paraná, in Curitiba.

Forestry Education Today

Prescribed burning is used by foresters to reduce fuel loads

Today, forestry education typically includes training in general biology, botany, genetics, soil science, climatology, hydrology, economics and forest management. Education in the basics of sociology and political science is often considered an advantage.

In India, forestry education is imparted in the agricultural universities and in Forest Research Institutes (deemed universities). Four year degree programmes are conducted in these universities at the undergraduate level. Masters and Doctorate degrees are also available in these universities.

In the United States, postsecondary forestry education leading to a Bachelor's degree or Master's degree is accredited by the Society of American Foresters.

In Canada the Canadian Institute of Forestry awards silver rings to graduates from accredited university BSc programs, as well as college and technical programs.

In many European countries, training in forestry is made in accordance with requirements of the Bologna Process and the European Higher Education Area.

The International Union of Forest Research Organizations is the only international organization that coordinates forest science efforts world-wide.

Forest

A conifer forest in the Swiss Alps (National Park)

The Adirondack Mountains of Upstate New York form the southernmost part of the Eastern forest-boreal transition ecoregion.

Forest on Mount Dajt, Albania

A forest is a large area of land covered with trees or other woody vegetation. Hundreds of more precise definitions of forest are used throughout the world, incorporating factors such as tree density, tree height, land use, legal standing and ecological function. According to the widely used United Nations Food and Agriculture Organization definition, forests covered four billion hectares (15 million square miles) or approximately 30 percent of the world's land area in 2006.

Forests are the dominant terrestrial ecosystem of Earth, and are distributed across the globe. Forests account for 75% of the gross primary productivity of the Earth's biosphere, and contain 80% of the Earth's plant biomass.

Forests at different latitudes form distinctly different ecozones: boreal forests near the poles tend to consist of evergreens, while tropical forests near the equator tend to be distinct from the temperate forests at mid-latitude. The amount of precipitation and the elevation of the forest also affects forest composition.

Human society and forests influence each other in both positive and negative ways. Forests provide ecosystem services to humans and serve as tourist attractions. Forests can also impose costs, affect people's health, and interfere with tourist enjoyment. Human activities, including harvesting forest resources, can negatively affect forest ecosystems.

Definition

Forest in the Scottish Highlands

Although *forest* is a term of common parlance, there is no universally recognised precise definition, with more than 800 definitions of forest used around the world. Although a forest is usually defined by the presence of trees, under many definitions an area completely lacking trees may still be considered a forest if it grew trees in the past, will grow trees in the future, or was legally designated as a forest regardless of vegetation type.

There are three broad categories of forest definitions in use: administrative, land use, and land cover. Administrative definitions are based primarily upon the legal designations of land, and commonly bear little relationship to the vegetation growing on the land: land that is legally designated as a forest is defined as a forest even if no trees are growing on it. Land use definitions are based upon the primary purpose that the land serves. For example, a forest may defined as any land that is used primarily for production of timber. Under such a land use definition, cleared roads or infrastructure within an area used for forestry, or areas within the region that have been cleared by harvesting, disease or fire are still considered forests even if they contain no trees. Land cover definitions define forests based upon the type and density of vegetation growing on the land. Such definitions typically define a forest as an area growing trees above some threshold. These thresholds are typically the number of trees per area (density), the area of ground under the tree canopy (canopy cover) or the section of land that is occupied by the cross-section of tree trunks (basal area). Under such land cover definitions, and area of land only be defined as forest if it is growing trees. Areas that fail to meet the land cover definition may be still included under while immature trees are establishing if they are expected to meet the definition at maturity.

Under land use definitions, there is considerable variation on where the cutoff points are between a forest, woodland, and savanna. Under some definitions, forests require very high levels of tree canopy cover, from 60% to 100%, excluding savannas and woodlands in which trees have a lower canopy cover. Other definitions consider savannas to be a type of forest, and include all areas with tree canopies over 10%.

Conversely some areas covered in trees are legally defined as agricultural areas. For example, Norway spruce plantations in Austrian forest law when the trees are being grown as Christmas trees and below a certain height.

Etymology

The word *forest* comes from Middle English, from Old French *forest* (also *forès*) "forest, vast expanse covered by trees"; first introduced in English as the word for wild land set aside for hunting without the necessity in definition for the existence of trees. Possibly a borrowing (probably via Frankish or Old High German) of the Medieval Latin word *foresta* "open wood", *foresta* was first used by Carolingian scribes in the Capitularies of Charlemagne to refer specifically to the king's royal hunting grounds. The term was not endemic to Romance languages (e.g. native words for "forest" in the Romance languages evolved out of the Latin word *silva* "forest, wood" (English *sylvan*); cf. Italian, Spanish, Portuguese *selva*; Romanian *silvă*; Old French *selve*); and cognates in Romance languages, such as Italian *foresta*, Spanish and Portuguese *floresta*, etc. are all ultimately borrowings of the French word.

Since the 13th century, the Niepołomice Forest in Poland has had special use and protection. In this view from space, different coloration can indicate different functions.

The exact origin of Medieval Latin *foresta* is obscure. Some authorities claim the word derives from the Late Latin phrase *forestam silvam*, meaning "the outer wood"; others claim the term is a latinisation of the Frankish word **forhist* "forest, wooded country", assimilated to *forestam silvam* (a common practice among Frankish scribes). Frankish **forhist* is attested by Old High German *forst* "forest", Middle Low German *vorst* "forest", Old English *fyrhþ* "forest, woodland, game preserve, hunting ground" (English *frith*), and Old Norse *fýri* "coniferous forest", all of which derive from Proto-Germanic **furhísa-*, **furhíþija-* "a fir-wood, coniferous forest", from +Proto-Indo-European **perkʷu-* "a coniferous or mountain forest, wooded height".

Uses of the word "forest" in English to denote any uninhabited area of non-enclosure are now considered archaic. The word was introduced by the Norman rulers of England as a legal term (appearing in Latin texts like the Magna Carta) denoting an uncultivated area legally set aside for hunting by feudal nobility.

Tywi forest, in Wales

These hunting forests were not necessarily wooded much, if at all. However, as hunting forests did often include considerable areas of woodland, the word "forest" eventually came to mean wooded land more generally. By the start of the 14th century, the word appeared in English texts, indicating all three senses: the most common one, the legal term and the archaic usage. Other terms used to mean "an area with a high density of trees" are *wood, woodland, wold, weald, holt, frith* and *firth*. Unlike *forest*, these are all derived from Old English and were not borrowed from another language. Some classifications now reserve the term *woodland* for an area with more open space between trees and distinguish among woodlands, *open forests*, and *closed forests* based on crown cover.

Evolution

The first known forests on Earth arose in the Late Devonian (approximately 380 million years ago), with the evolution of *Archaeopteris*. *Archaeopteris* was a plant that was both tree-like and fern-like, growing to 10 metres (33 ft) in height. *Archaeopteris* quickly spread throughout the world, from the equator to subpolar latitudes. *Archaeopteris* formed the first forest by being the first known species to cast shade due to its fronds and forming soil from its roots. *Archaeopteris* was deciduous, dropping its fronds onto the forest floor. The shade, soil, and forest duff from the dropped fronds created the first forest. The shed organic matter altered the freshwater environment, slowing it down and providing food. This promoted freshwater fish.

Ecology

Temperate rainforest in Tasmania's Hellyer Gorge

Forests account for 75% of the gross primary productivity of the Earth's biosphere, and contain 80% of the Earth's plant biomass. Forest ecosystems can be found in all regions capable of sustaining tree growth, at altitudes up to the tree line, except where natural fire frequency or other disturbance is too high, or where the environment has been altered by human activity.

The latitudes 10° north and south of the equator are mostly covered in tropical rainforest, and the latitudes between 53°N and 67°N have boreal forest. As a general rule, forests dominated by angiosperms (*broadleaf forests*) are more species-rich than those dominated by gymnosperms (*conifer*, *montane*, or *needleleaf forests*), although exceptions exist.

Forests sometimes contain many tree species within a small area (as in tropical rain and temperate deciduous forests), or relatively few species over large areas (e.g., taiga and arid montane coniferous forests). Forests are often home to many animal and plant species, and biomass per unit area is high compared to other vegetation communities. Much of this biomass occurs below ground in the root systems and as partially decomposed plant detritus. The woody component of a forest contains lignin, which is relatively slow to decompose compared with other organic materials such as cellulose or carbohydrate.

Components

Even, dense old-growth stand of beech trees (*Fagus sylvatica*) prepared to be regenerated by their saplings in the understory, in the Brussels part of the Sonian Forest.

A forest consists of many components that can be broadly divided into two categories that are biotic (living) and abiotic (non-living) components. The living parts include trees, shrubs, vines, grasses and other herbaceous (non-woody) plants, mosses, algae, fungi, insects, mammals, birds, reptiles, amphibians, and microorganisms living on the plants and animals and in the soil.

Layers

Biogradska forest in Montenegro

Spiny forest at Ifaty, Madagascar, featuring various *Adansonia* (baobab) species, *Alluaudia procera* (Madagascar ocotillo) and other vegetation

A forest is made up of many layers. Starting from the ground level and moving up, the main layers of all forest types are the forest floor, the understory and the canopy. The emergent layer exists in tropical rainforests. Each layer has a different set of plants and animals depending upon the availability of sunlight, moisture and food.

- Forest floor contains decomposing leaves, animal droppings, and dead trees. All of decay on the forest floor and form the new soil and provides nutrients to the plants. The forest floor supports ferns, grasses, mushroom and trees seedlings.
- Understory is made up of bushes, shrubs, and young trees that are adapted to living in the shades of the canopy.
- Canopy is formed by the mass of intertwined branches, twigs and leaves of the mature trees. The crowns of the dominant trees receive most of the sunlight. This is the most productive part of the trees where maximum food is produced. The canopy forms a shady, protective "umbrella" over the rest of the forest.
- Emergent layer exists in the tropical rain forest and is composed of a few scattered trees that tower over the canopy.

Types

Forests can be classified in different ways and to different degrees of specificity. One such way is in terms of the biome in which they exist, combined with leaf longevity of the dominant species (whether they are evergreen or deciduous). Another distinction is whether the forests are composed predominantly of broadleaf trees, coniferous (needle-leaved) trees, or mixed.

- Boreal forests occupy the subarctic zone and are generally evergreen and coniferous.
- Temperate zones support both broadleaf deciduous forests (*e.g.*, temperate deciduous forest) and evergreen coniferous forests (*e.g.*, temperate coniferous

forests and temperate rainforests). Warm temperate zones support broadleaf evergreen forests, including laurel forests.

- Tropical and subtropical forests include tropical and subtropical moist forests, tropical and subtropical dry forests, and tropical and subtropical coniferous forests.
- Physiognomy classifies forests based on their overall physical structure or developmental stage (e.g. old growth vs. second growth).
- Forests can also be classified more specifically based on the climate and the dominant tree species present, resulting in numerous different forest types (e.g., Ponderosa pine/Douglas-fir forest).

The number of trees in the world, according to a 2015 estimate, is 3 trillion, of which 1.4 trillion are in the tropics or sub-tropics, 0.6 trillion in the temperate zones, and 0.7 trillion in the coniferous boreal forests. The estimate is about eight times higher than previous estimates, and is based on tree densities measured on over 400,000 plots. It remains subject to a wide margin of error, not least because the samples are mainly from Europe and North America.

Forests can also be classified according to the amount of human alteration. Old-growth forest contains mainly natural patterns of biodiversity in established seral patterns, and they contain mainly species native to the region and habitat. In contrast, secondary forest often contains significant elements of species which were originally from other regions or habitats.

Different global forest classification systems have been proposed, but none has gained universal acceptance. UNEP-WCMC's forest category classification system is a simplification of other more complex systems (e.g. UNESCO's forest and woodland 'subformations'). This system divides the world's forests into 26 major types, which reflect climatic zones as well as the principal types of trees. These 26 major types can be reclassified into 6 broader categories: temperate needleleaf; temperate broadleaf and mixed; tropical moist; tropical dry; sparse trees and parkland; and forest plantations. Each category is described as a separate section below.

Temperate Needleleaf

Temperate needleleaf forests mostly occupy the higher latitude regions of the northern hemisphere, as well as high altitude zones and some warm temperate areas, especially on nutrient-poor or otherwise unfavourable soils. These forests are composed entirely, or nearly so, of coniferous species (Coniferophyta). In the Northern Hemisphere pines *Pinus*, spruces *Picea*, larches *Larix*, firs *Abies*, Douglas firs *Pseudotsuga* and hemlocks *Tsuga*, make up the canopy, but other taxa are also important. In the Southern Hemisphere, most coniferous trees (members of the Araucariaceae and Podocarpaceae) occur in mixtures with broadleaf species, and are classed as broadleaf and mixed forests.

Temperate Broadleaf and Mixed

Broadleaf forest in Bhutan

Temperate broadleaf and mixed forests include a substantial component of trees in the Anthophyta. They are generally characteristic of the warmer temperate latitudes, but extend to cool temperate ones, particularly in the southern hemisphere. They include such forest types as the mixed deciduous forests of the United States and their counterparts in China and Japan, the broadleaf evergreen rainforests of Japan, Chile and Tasmania, the sclerophyllous forests of Australia, central Chile, the Mediterranean and California, and the southern beech Nothofagus forests of Chile and New Zealand.

Tropical Moist

There are many different types of tropical moist forests, with lowland evergreen broad leaf tropical rainforests, for example várzea and igapó forests and the terra firma forests of the Amazon Basin; the peat swamp forests, dipterocarp forests of Southeast Asia; and the high forests of the Congo Basin. Seasonal tropical forests, perhaps the best description for the colloquial term "jungle", typically range from the rainforest zone 10 degrees north or south of the equator, to the Tropic of Cancer and Tropic of Capricorn. Forests located on mountains are also included in this category, divided largely into upper and lower montane formations on the basis of the variation of physiognomy corresponding to changes in altitude.

Tropical Dry

Tropical dry forests are characteristic of areas in the tropics affected by seasonal drought. The seasonality of rainfall is usually reflected in the deciduousness of the forest canopy, with most trees being leafless for several months of the year. However, under some conditions, e.g. less fertile soils or less predictable drought regimes, the proportion of evergreen species increases and the forests are characterised as "sclerophyllous". Thorn forest, a dense forest of low stature with a high frequency of thorny or spiny species, is found where drought is prolonged, and especially where grazing animals are plentiful. On very poor soils, and especially where fire is a recurrent phenomenon, woody savannas develop.

Sparse Trees and Parkland

Taiga forest near Saranpaul in the northeast Ural Mountains, Khanty–Mansia, Russia. Trees include *Picea obovata* (dominant on right bank), *Larix sibirica, Pinus sibirica,* and *Betula pendula.*

Sparse trees and parkland are forests with open canopies of 10–30% crown cover. They occur principally in areas of transition from forested to non-forested landscapes. The two major zones in which these ecosystems occur are in the boreal region and in the seasonally dry tropics. At high latitudes, north of the main zone of boreal forest or taiga, growing conditions are not adequate to maintain a continuous closed forest cover, so tree cover is both sparse and discontinuous. This vegetation is variously called open taiga, open lichen woodland, and forest tundra. It is species-poor, has high bryophyte cover, and is frequently affected by fire.

Forest Plantations

Forest plantations, generally intended for the production of timber and pulpwood increase the total area of forest worldwide. Commonly mono-specific and/or composed of introduced tree species, these ecosystems are not generally important as habitat for native biodiversity. However, they can be managed in ways that enhance their biodiversity protection functions and they are important providers of ecosystem services such as maintaining nutrient capital, protecting watersheds and soil structure as well as storing carbon. They may also play an important role in alleviating pressure on natural forests for timber and fuelwood production.

Societal Significance

Forests may provide a diversity of ecosystem services including recycling carbon dioxide into oxygen, acting as a carbon sink, aiding in regulating climate, purify water, mitigating natural hazards such as floods, and serving as a genetic reserve. Forests also serve as a source of lumber and as recreational areas.

Coastal Douglas fir woodland in northwest Oregon

Redwood tree in northern California redwood forest, where many redwood trees are managed for preservation and longevity, rather than being harvested for wood production

A forest near Vinitsa, Republic of Macedonia

Some researchers state that forests do not only provide benefits, but can also incur large and significant costs to humans. Forests may impose a considerable economic burden, diminish the enjoyment of natural areas, cause a dramatic decline in the food producing capacity of grazing land and cultivated land, reduce biodiversity remove available water for humans and wildlife, harbour dangerous and destructive wildlife, and act as reservoirs of human and livestock disease.

The management of forests is often referred to as forestry. Forest management has changed considerably over the last few centuries, with rapid changes from the 1980s onwards culminating in a practice now referred to as sustainable forest management.

Forest ecologists concentrate on forest patterns and processes, usually with the aim of elucidating cause-and-effect relationships. Foresters who practice sustainable forest management focus on the integration of ecological, social, and economic values, often in consultation with local communities and other stakeholders.

Humans have generally decreased the amount of forest worldwide. Anthropogenic factors that can affect forests include logging, urban sprawl, human-caused forest fires, acid rain, invasive species, and the slash and burn practices of swidden agriculture or shifting cultivation. The loss and re-growth of forest leads to a distinction between two broad types of forest, primary or old-growth forest and secondary forest. There are also many natural factors that can cause changes in forests over time including forest fires, insects, diseases, weather, competition between species, etc. In 1997, the World Resources Institute recorded that only 20% of the world's original forests remained in large intact tracts of undisturbed forest. More than 75% of these intact forests lie in three countries—the boreal forests of Russia and Canada and the rainforest of Brazil.

In 2010, the Food and Agriculture Organization of the United Nations reported that world deforestation, mainly the conversion of tropical forests to agricultural land, had decreased over the past ten years but still continues at a high rate in many countries. Globally, around 13 million hectares of forests were converted to other uses or lost through natural causes each year between 2000 and 2010 as compared to around 16 million hectares per year during the 1990s. The study covered 233 countries and areas. Brazil and Indonesia, which had the highest loss of forests in the 1990s, have significantly reduced their deforestation rates. China instituted a ban on logging, beginning in 1998, due to the erosion and flooding that it caused. In addition, ambitious tree planting programmes in countries such as China, India, the United States and Vietnam - combined with natural expansion of forests in some regions - have added more than seven million hectares of new forests annually. As a result, the net loss of forest area was reduced to 5.2 million hectares per year between 2000 and 2010, down from 8.3 million hectares annually in the 1990s. In 2015, a study for *Nature Climate Change* showed that the trend has recently been reversed, leading to an "overall gain" in global biomass and forests. This gain is due especially to reforestation in China and Russia. However new forests are not completely equivalent to old growth forests in terms of species diversity, resilience and carbon capture. On September 7, 2015, the Food and Agriculture Organization of the United Nations released a new study stating that, over the last 25 years, the global deforestation rate has decreased by 50% due to improved management of forests and greater government protection.

Smaller areas of woodland in cities may be managed as urban forestry, sometimes within public parks. These are often created for human benefits; Attention Restoration Theory argues that spending time in nature reduces stress and improves health, while forest schools and kindergartens help young people to develop social as well as scientific skills in forests. These typically need to be close to where the children live, for practical logistics.

Canada

Garibaldi Provincial Park, British Columbia

Canada has about 4,020,000 square kilometres (1,550,000 sq mi) of forest land. More than 90% of forest land is publicly owned and about 50% of the total forest area is allocated for harvesting. These allocated areas are managed using the principles of sustainable forest management, which includes extensive consultation with local stakeholders. About eight percent of Canada's forest is legally protected from resource development. Much more forest land—about 40 percent of the total forest land base—is subject to varying degrees of protection through processes such as integrated land use planning or defined management areas such as certified forests.

By December 2006, over 1,237,000 square kilometers of forest land in Canada (about half the global total) had been certified as being sustainably managed. Clearcutting, first used in the latter half of the 20th century, is less expensive, but devastating to the environment, and companies are required by law to ensure that harvested areas are adequately regenerated. Most Canadian provinces have regulations limiting the size of clearcuts, although some older clearcuts can range upwards of 110 square kilometres (27,000 acres) in size which were cut over several years.

United States

Trees on a mountain in northern Utah during early autumn.

In the United States, most forests have historically been affected by humans to some degree, though in recent years improved forestry practices has helped regulate or moderate large scale or severe impacts. However, the United States Forest Service estimates a net loss of about 2 million hectares (4,942,000 acres) between 1997 and 2020; this estimate includes conversion of forest land to other uses, including urban and suburban development, as well as afforestation and natural reversion of abandoned crop and pasture land to forest. However, in many areas of the United States, the area of forest is stable or increasing, particularly in many northern states. The opposite problem from flooding has plagued national forests, with loggers complaining that a lack of thinning and proper forest management has resulted in large forest fires.

Land Area

Forest land area										
	2008		2009		2010		2011		2012	
	('000 km²)	('000 mi²)	('000 km²)	('000 mi²)	('000 km²)	('000 mi²)	('000 km²)	('000 mi²)	('000. km²)	('000 mi²)
Australia	1,511	583	1,502	580	1,493	576	1,473	569	1,467	566
Brazil	5,239	2,023	5,217	2,014	5,195	2,006	5,183	2,001	5,173	1,997
Canada	3,101	1,197	3,101	1,197	3,101	1,197	3,101	1,197	3,101	1,197
China	2,013	777	2,041	788	2,069	799	2,159	834	2,168	837
European Union	1,559	602	1,564	604	1,569	606	1,573	607	1,578	609
Germany	111	43	111	43	111	43	113	44	113	44
India	681	263	683	264	684	264	693	268	693	268
Indonesia	958	370	951	367	944	364	937	362	931	359
Japan	250	97	250	97	250	97	251	97	251	97
Russia	8,090	3,120	8,090	3,120	8,091	3,124	8,092	3,124	8,093	3,125
United States	3,033	1,171	3,036	1,172	3,040	1,170	3,049	1,177	3,051	1,178
World total	40,318	15,567	40,261	15,545	40,204	15,523	40,184	15,515	39,519	15,258

References

- T. Mirov, Nicholas; Hasbrouck, Jean (1976). "6". The story of pines. Bloomington and London: Indiana University Press. p. 111. ISBN 0-253-35462-5.
- Braudel, Fernand (1979). The Wheels of Commerce: Civilization and Capitalism: 15th-18th Century (Volume II). University of California Press. p. 240. ISBN 978-0-520-08115-4.
- Johnson, F.X.; Pacini, H.; Smeets, E (2013). Transformations in EU biofuels markets under the Renewable Energy Directive and the implications for land use, trade and forests. CIFOR. p. 32. ISBN 9786028693813.
- Ritter, E (2011). Dauksta, D, ed. New Perspectives on People and Forests. World Forests. 9. Springer. ISBN 94-007-1149-2.
- "forest, noun". The American Heritage Dictionary of the English Language (3 ed.). Boston: Houghton Mifflin Company. 1996. ISBN 0-395-44895-6.
- Jenkins, Martin D.; Groombridge, Brian (2002). World Atlas of Biodiversity: Earth's Living Resources in the 21st Century. World Conservation Monitoring Centre, United Nations Environment Programme. ISBN 9780520236684.

- Chape, S; Spalding, M; Jenkins, M (2008). The world's protected areas: status, values and prospects in the 21st century. Univ de Castilla La Mancha. ISBN 0520246608.
- Wilcox, B.A.; Ellis, B. "Forests and emerging infectious diseases of humans". Food and Agriculture Organization of the United Nations. Retrieved 2014-11-16.
- Schuck, Andreas; Päivinen, Risto; Hytönend, Tuomo; Pajari, Brita (2002). "Compilation of Forestry Terms and Definitions" (PDF). Joensuu, Finland: European Forest Institute. Retrieved 2014-11-16.
- "Definitions: Indicative definitions taken from the Report of the ad hoc technical expert group on forest biological diversity". Convention on Biological Diversity. Retrieved 2014-11-16.
- MacDicken, Kenneth (2013-03-15). "Forest Resources Assessment Working Paper 180" (PDF). Rome: Food and Agriculture Organization of the United Nations Forestry Department. Retrieved 2014-11-16.
- Watson, Robert T.; Verardo, David J.; Noble, Ian R.; Bolin, Bert; Ravindranath, N.H.; Dokken, David J., eds. (2000). "Land Use, Land-Use Change and Forestry". Intergovernmental Panel on Climate Change. Retrieved 2014-11-16.
- "Comparative framework and Options for harmonization of definitions". Second Expert Meeting on Harmonizing Forest-Related Definitions. FAO. Retrieved 2014-11-26.
- "FAO. 2010. Global Forest Resources Assessment 2010. Main report. FAO Forestry Paper 163. Rome, Italy". Fao.org. Retrieved 3 July 2013.
- Petru-Ioan Becheru (Aug 2012). "Revista pădurilor online". Rev. pădur. (in Romanian). 127 (4): 46–53. ISSN 1583-7890. 16819. Retrieved 2012-10-21.

2

Types of Forests

Forests that receive more than 2000 mm of rainfall are known as tropical evergreen forests. The rainfall in rainforests varies from 250 to 450 centimeters. The other types of forests mentioned in this text are temperate coniferous forest, temperate deciduous forest, freshwater swamp forest and tropical and sub tropical moist broadleaf forests. Forests can be best understood in confluence with the major types listed in the following section.

Evergreen Forest

An evergreen forest is a forest consisting entirely or mainly of evergreen trees that retain green foliage all year round. Such forests may be broadleaved evergreen forests, such as *Moist forest* that occur in tropical climates or coniferous evergreen forests such as those found in temperate and boreal climates. Some subtropical or warm temperate areas may contain *laurel forest*, which are broadleaved evergreen, while isolated areas of evergreen forests may occur in *montane forests,* and *cloud forests*.

The Tropical Evergreen Forests

Tropical evergreen forests or (tropical rain forests) are usually found in areas receiving more than 200 mm of rainfall and having a monthly mean temperature of 18 °C or higher in the coldest months. They occupy about seven per cent the Earth's land surface and harbour more than half of the planet's terrestrial plants and animals. Tropical evergreen forests are dense, multi-layered, and harbour many types of plants and animals. The trees are evergreen as there is no period of drought or frost. The canopy tree species are mostly tall hardwoods with broad leaves that release large quantities of water through transpiration, in a cycle that is important in raising as much mineral nutrient material as possible from the soil.

Example:

In India, evergreen forests are found on the eastern and western slopes of the Western Ghats in such states as Tamil Nadu, Karnataka, Kerala and Maharashtra. And also found in Assam, Arunachal Pradesh, Meghalaya, Nagaland, Tripura, West Bengal and Andaman and Nicobar Islands. They are also found in the hills of Jaintia and Khasi. Some of the trees found in Indian tropical forests are rosewood, mahogany and ebony. Bamboo and reeds are also common. Because of dense foliage competing for light, little direct sunlight reaches the under story.

Temperate evergreen forests, coniferous, broadleaf, and mixed, are found largely in the temperate mid-latitudes of North America, Siberia, Canada, Australia, Africa and Scandinavia. Broadleaf evergreen forests occur in particular in southern China, southeastern Brazil, parts of southeastern North America, and in countries around the Mediterranean Basin, such as Lebanon and Morocco. Other examples include the wet temperate conifer forests of northwestern North America.

Coniferous temperate evergreen forests are most frequently dominated by species in the families Pinaceae and Cupressaceae. Broadleaf temperate evergreen forests include those in which Fagaceae such as oaks are common, those in which Nothofagaceae predominate, and the Eucalyptus forests of the Southern Hemisphere. There also are assorted temperate evergreen forests dominated by other families of trees, such as Lauraceae in laurel forest.

Evergreen forests around the world are under threat of logging, mining, oil and gas developments, pollution, hydroelectric projects and other human developments planned in these areas.

Rainforest

The Daintree Rainforest in Queensland, Australia

The Daintree Rainforest near Cairns, in Queensland, Australia

Rainforests are forests characterized by high rainfall, with annual rainfall between 250 and 450 centimetres (98 and 177 in). There are two types of rainforest: tropical rainforest and temperate rainforest. The monsoon trough, alternatively known as the intertropical convergence zone, plays a significant role in creating the climatic conditions necessary for the Earth's tropical rainforests.

Around 40% to 75% of all biotic species are indigenous to the rainforests. It has been estimated that there may be many millions of species of plants, insects and microorganisms still undiscovered in tropical rainforests. Tropical rainforests have been called the "jewels of the Earth" and the "world's largest pharmacy", because over one quarter of natural medicines have been discovered there. Rainforests are also responsible for 28% of the world's oxygen turnover, sometimes misnamed oxygen production, processing it through photosynthesis from carbon dioxide and consuming it through respiration.

The undergrowth in some areas of a rainforest can be restricted by poor penetration of sunlight to ground level. If the leaf canopy is destroyed or thinned, the ground beneath is soon colonized by a dense, tangled growth of vines, shrubs and small trees, called a jungle. The term *jungle* is also sometimes applied to tropical rainforests generally.

Tropical

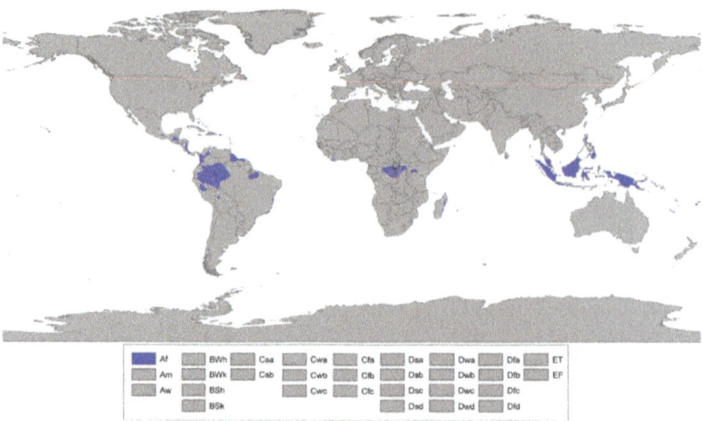

Worldwide tropical rainforest zones.

Tropical rainforests are characterized by a warm and wet climate with no substantial dry season: typically found within 10 degrees north and south of the equator. Mean monthly temperatures exceed 18 °C (64 °F) during all months of the year. Average annual rainfall is no less than 168 cm (66 in) and can exceed 1,000 cm (390 in) although it typically lies between 175 cm (69 in) and 200 cm (79 in).

Many of the world's tropical forests are associated with the location of the monsoon trough, also known as the intertropical convergence zone. The broader category of tropical moist forests are located in the equatorial zone between the Tropic of Cancer and Tropic of Capricorn. Tropical rainforests exist in Southeast Asia (from Myanmar (Burma) to the Philippines, Malaysia, Indonesia, Papua New Guinea, Sri Lanka, Sub-Saharan Africa from Cameroon to the Congo (Congo Rainforest), South America (e.g. the Amazon Rainforest), Central America (e.g. Bosawás, southern Yucatán Peninsula-El Peten-Belize-Calakmul), Australia, and on many of the Pacific Islands (such as Hawai□i). Tropical forests have been called the "Earth's lungs", although it is now known that rainforests contribute little net oxygen addition to the atmosphere through photosynthesis.

Temperate

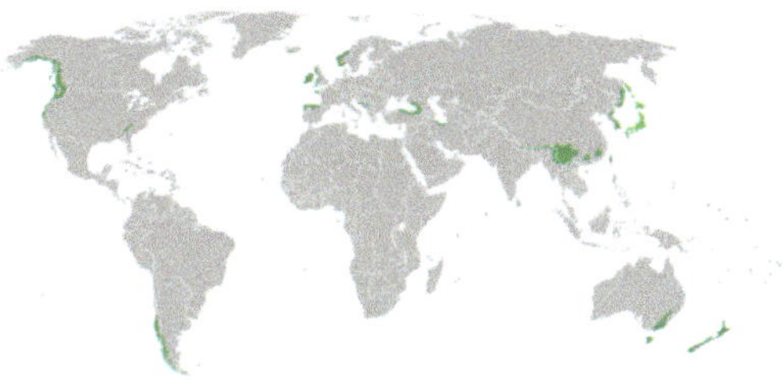

General distribution of temperate rainforests

Tropical forests cover a large part of the globe, but temperate rainforests only occur in few regions around the world. Temperate rainforests are rainforests in temperate regions. They occur in North America (in the Pacific Northwest in Alaska, British Columbia, Washington, Oregon and California), in Europe (parts of the British Isles such as the coastal areas of Ireland and Scotland, southern Norway, parts of the western Balkans along the Adriatic coast, as well as in Galicia and coastal areas of the eastern Black Sea, including Georgia and coastal Turkey), in East Asia (in southern China, Highlands of Taiwan, much of Japan and Korea, and on Sakhalin Island and the adjacent Russian Far East coast), in South America (southern Chile) and also in Australia and New Zealand.

Layers

A tropical rainforest typically has a number of layers, each with different plants and animals adapted for life in that particular area. Examples include the emergent, canopy, understorey and forest floor layers.

Emergent Layer

The emergent layer contains a small number of very large trees called emergents, which grow above the general canopy, reaching heights of 45–55 m, although on occasion a few species will grow to 70–80 m tall. They need to be able to withstand the hot temperatures and strong winds that occur above the canopy in some areas. Eagles, butterflies, bats and certain monkeys inhabit this layer.

The canopy at the Forest Research Institute Malaysia showing crown shyness

Canopy Layer

The canopy layer contains the majority of the largest trees, typically 30 metres (98 ft) to 45 metres (148 ft) tall. The densest areas of biodiversity are found in the forest canopy, a more or less continuous cover of foliage formed by adjacent treetops. The canopy, by some estimates, is home to 50 percent of all plant species. Epiphytic plants attach to trunks and branches, and obtain water and minerals from rain and debris that collects on the supporting plants. The fauna is similar to that found in the emergent layer, but more diverse. A quarter of all insect species are believed to exist in the rainforest canopy. Scientists have long suspected the richness of the canopy as a habitat, but have only recently developed practical methods of exploring it. As long ago as 1917, naturalist William Beebe declared that "another continent of life remains to be discovered, not upon the Earth, but one to two hundred feet above it, extending over thousands of square miles." True exploration of this habitat only began in the 1980s, when scientists developed methods to reach the canopy, such as firing ropes into the trees using crossbows. Exploration of the canopy is still in its infancy, but other methods include the use of balloons and airships to float above the highest branches and the building of cranes and walkways planted on the forest floor. The science of accessing tropical forest canopy using airships or similar aerial platforms is called dendronautics.

Understory Layer

The understory or understorey layer lies between the canopy and the forest floor. It is home to a number of birds, snakes and lizards, as well as predators such as jaguars, boa constrictors and leopards. The leaves are much larger at this level and insect life is abundant. Many seedlings that will grow to the canopy level are present in the understory. Only about 5% of the sunlight shining on the rainforest canopy reaches the understory. This layer can be called a *shrub layer*, although the shrub layer may also be considered a separate layer.

Forest Floor

Rainforest in the Blue Mountains, Australia

The forest floor, the bottom-most layer, receives only 2% of the sunlight. Only plants adapted to low light can grow in this region. Away from riverbanks, swamps and clearings, where dense undergrowth is found, the forest floor is relatively clear of vegetation because of the low sunlight penetration. It also contains decaying plant and animal matter, which disappears quickly, because the warm, humid conditions promote rapid decay. Many forms of fungi growing here help decay the animal and plant waste.

Flora and Fauna

West Usambara Two-Horned Chameleon (*Bradypodion fischeri*) in the Usambara Mountains, Tanzania.

More than half of the world's species of plants and animals are found in the rainforest. Rainforests support a very broad array of fauna, including mammals, reptiles, birds and invertebrates. Mammals may include primates, felids and other families. Reptiles include snakes, turtles, chameleons and other families; while birds include such families as vangidae and Cuculidae. Dozens of families of invertebrates are found in rainforests. Fungi are also very common in rainforest areas as they can feed on the decomposing remains of plants and animals. Many rainforest species are rapidly disappearing due to deforestation, habitat loss and pollution of the atmosphere.

Soils

Despite the growth of vegetation in a tropical rainforest, soil quality is often quite poor. Rapid bacterial decay prevents the accumulation of humus. The concentration of iron and aluminium oxides by the laterization process gives the oxisols a bright red colour and sometimes produces mineral deposits such as bauxite. Most trees have roots near the surface, because there are insufficient nutrients below the surface; most of the trees' minerals come from the top layer of decomposing leaves and animals. On younger substrates, especially of volcanic origin, tropical soils may be quite fertile. If rainforest trees are cleared, rain can accumulate on the exposed soil surfaces, creating run-off and beginning a process of soil erosion. Eventually streams and rivers form and flooding becomes possible.

Effect on global Climate

A natural rainforest emits and absorbs vast quantities of carbon dioxide. On a global scale, long-term fluxes are approximately in balance, so that an undisturbed rainforest would have a small net impact on atmospheric carbon dioxide levels, though they may have other climatic effects (on cloud formation, for example, by recycling water vapour). No rainforest today can be considered to be undisturbed. Human-induced deforestation plays a significant role in causing rainforests to release carbon dioxide, as do other factors, whether human-induced or natural, which result in tree death, such as burning and drought. Some climate models operating with interactive vegetation predict a large loss of Amazonian rainforest around 2050 due to drought, forest dieback and the subsequent release more carbon dioxide. Five million years from now, the Amazon rainforest may long since have dried and transformed itself into savannah, killing itself in the progress (changes such as this may happen even if all human deforestation activity ceases overnight). The descendants of our known animals may adapt to the dry savannah of the former Amazonian rainforest and thrive in the new, warmer temperatures.

Human Uses

Aerial view of the Amazon rainforest, taken from a plane.

Tropical rainforests provide timber as well as animal products such as meat and hides. Rainforests also have value as tourism destinations and for the ecosystem services provided. Many foods originally came from tropical forests, and are still mostly grown on plantations in regions that were formerly primary forest. Also, plant-derived medicines are commonly used for fever, fungal infections, burns, gastrointestinal problems, pain, respiratory problems, and wound treatment.

Native Peoples

On January 18, 2007, FUNAI reported also that it had confirmed the presence of 67 different uncontacted tribes in Brazil, up from 40 in 2005. With this addition, Brazil has now overtaken the island of New Guinea as the country having the largest number of uncontacted tribes. The province of Irian Jaya or West Papua in the island of New Guinea is home to an estimated 44 uncontacted tribal groups. The tribes are in danger because of the deforestation, especially in Brazil.

Central African rainforest is home of the Mbuti pygmies, one of the hunter-gatherer peoples living in equatorial rainforests characterised by their short height (below one and a half metres, or 59 inches, on average). They were the subject of a study by Colin Turnbull, *The Forest People*, in 1962. Pygmies who live in Southeast Asia are, amongst others, referred to as "Negrito".

Deforestation

Satellite photograph of the haze above Borneo and Sumatra on 24 September 2015.

Tropical and temperate rainforests have been subjected to heavy logging and agricultural clearance throughout the 20th century and the area covered by rainforests around the world is shrinking. Biologists have estimated that large numbers of species are being driven to extinction (possibly more than 50,000 a year; at that rate, says E. O. Wilson of Harvard University, a quarter or more of all species on Earth could be exterminated within 50 years) due to the removal of habitat with destruction of the rainforests.

Another factor causing the loss of rainforest is expanding urban areas. Littoral rainforest growing along coastal areas of eastern Australia is now rare due to ribbon development to accommodate the demand for seachange lifestyles.

The forests are being destroyed at a rapid pace. Almost 90% of West Africa's rainforest has been destroyed. Since the arrival of humans, Madagascar has lost two thirds of its original rainforest. At present rates, tropical rainforests in Indonesia would be logged out in 10 years and Papua New Guinea in 13 to 16 years. According to Rainforest Rescue, a main reason for the increasing deforestation rate especially in Indonesia is the expansion of oil palm plantations to meet the growing demand for cheap vegetable fats and biofuels. In Indonesia, palm oil is already cultivated on nine million hectares and, together with Malaysia, the island nation produces about 85 percent of the world's palm oil.

Several countries, notably Brazil, have declared their deforestation a national emergency. Amazon deforestation jumped by 69% in 2008 compared to 2007's twelve months, according to official government data. Deforestation could wipe out or severely damage nearly 60% of the Amazon Rainforest by 2030, says a 2007 report from WWF.

However, a January 30, 2009 New York Times article stated, "By one estimate, for every acre of rain forest cut down each year, more than 50 acres of new forest are growing in the tropics..." The new forest includes secondary forest on former farmland and so-called degraded forest.

Temperate Coniferous Forest

A pine forest is an example of a temperate coniferous forest

Temperate coniferous forest is a terrestrial biome found in temperate regions of the world with warm summers and cool winters and adequate rainfall to sustain a forest. In most temperate coniferous forests, evergreen conifers predominate, while some are a mix of conifers and broadleaf evergreen trees and/or broadleaf deciduous trees. Temperate evergreen forests are common in the coastal areas of regions that have mild winters and heavy rainfall, or inland in drier climates or mountain areas. Coniferous forests can be found in the United States, Canada, Europe, and Asia. Many species of tree inhabit these forests including cedar, cypress, Douglas fir, fir, juniper, pine, podocarpus, spruce, redwood and yew. The understory also contains a wide variety of herbaceous and shrub species.

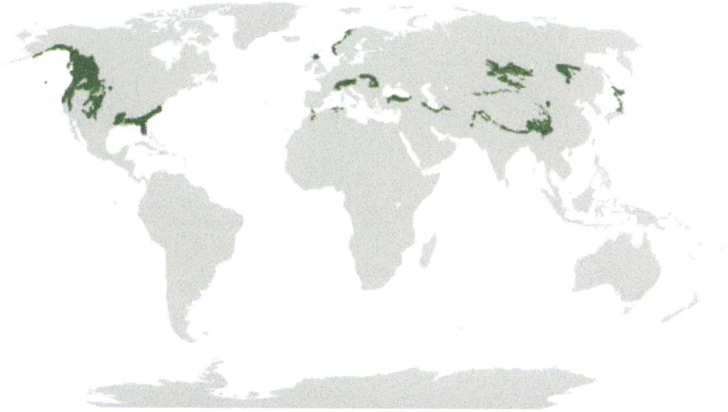

Map of temperate coniferous forest ecoregions of the world.

Structurally, these forests are rather simple, generally consisting of two layers: an overstory and understory. Some forests may support an intermediate layer of shrubs. Pine forests support a herbaceous understory that is generally dominated by grasses and herbaceous perennials, and are often subject to ecologically important wildfires.

Temperate rain forests occur only in seven regions around the world: the Pacific temperate rain forests of the Pacific Northwest, the Valdivian temperate rain forests of southwestern South America, the rain forests of New Zealand and Tasmania, northwest Europe (small pockets in Ireland, Scotland, Wales, Iceland and a somewhat larger area in Norway), southern Japan, and the eastern Black Sea-Caspian Sea region of Turkey and Georgia to northern Iran. The moist conditions of temperate rain forests generally support an understory of mosses, ferns and some shrubs. Temperate rain forests can be temperate coniferous forests or temperate broadleaf and mixed forests.

Species

The temperate coniferous rain forests around the world sustain the highest levels of biomass in any terrestrial ecosystem and are notable for trees of massive proportions, including coast redwood (*Sequoia sempervirens*), Douglas fir (*Pseudotsuga menziesii*), Sitka spruce (*Picea sitchensis*), Alerce (*Fitzroya cupressoides*) and kauri (*Agathis australis*). These forests are quite rare, occurring in small areas of North America, southwestern South America and northern New Zealand. The Klamath-Siskiyou forests of northwestern California and southwestern Oregon is known for its rich variety of plant and animal species, including many endemic species.

The leaves on a coniferous tree are needle like. The leaves have a waxy coating to help the leaves from being damaged.

Temperate Coniferous Forest Ecoregions

Eurasia

Carpathian montane conifer forest, Slovakia.

Cedrus deodara in the Western Himalayan subalpine conifer forests.

North America

A temperate coniferous forest ecosystem in Petroglyphs Provincial Park, Ontario.

South America

Fitzroya forest at Alerce Costero National Park, Chile.

Bor

In Russia, a coniferous forest, specifically pine or spruce forest is known as bor (Russian: бор). Bors typically grow on well-drained sandy soils, sandy clay soils and mild loams. The names of several Russian historical buildings and localities refer to the word "bor" as the original place where they were built, such as Borovitskaya Tower or Church of the Savior on Bor.

Temperate Deciduous Forest

Temperate forest in Germany just as the leaf canopy is opening

Temperate deciduous forests or temperate broad-leaf forests are dominated by trees that lose their leaves each year. They are found in areas with warm, moist summers and mild winters. The three major areas of this forest type occur in the Northern Hemisphere: eastern North America, East Asia, and Europe. Smaller areas occur in Australasia and southern South America. Examples of typical trees in the Northern Hemisphere's deciduous forests include oak, maple, beech, and elm, while in Southern Hemisphere, trees of the genus *Nothofagus* dominate this type of forest. The diversity of tree species is higher in regions where the winter is milder, and also in mountainous regions that provide an array of soil types and microclimates. One of the world's great protected examples of this forest type is found in Great Smoky Mountains National Park.

Ecology

Bloodroot, like many other spring ephemerals, flowers in the spring before the forest canopy appears

The principal factor operating in these forests is the seasonal appearance and disappearance of the canopy. Shade from the canopy limits the growth of many kinds of plants. Many species that are typical of these forests time their growth and flowering to the short period just before the canopy opens; hence they are known as spring ephemerals. Examples include trilliums and bloodroot. Most spring ephemerals are insect-pollinated, and the seeds themselves are often transported by ants, a mode of dispersal known as myrmecochory. There are a smaller number of species able to grow under the canopy, and even a few that grow during the period when leaves are being lost.

Many understory plants have leaf adaptions to cope with low light levels, and the need to exploit moving flecks of light on the forest floor. A few, such as Indian pipe and Corallorhiza orchids, have adapted to the shade by parasitism. The trees similarly are controlled by shade. Most tree seedlings require small gaps when trees fall and make space in order to regenerate. A few require larger gaps such as those produced by windstorms. Gradients of soil moisture, soil depth, elevation and aspect control the distribution of many trees, shrubs and herbaceous species. Some require unusual conditions such as steep slopes, infertile soil, and drought to escape competition from the more common tree species.

Pileated woodpeckers depend upon dead or dying trees as a source of food and for constructing their nests.

Many migratory birds time their arrival to coincide with the opening of the canopy, which provides the insects that are their principal food sources for raising young. Owing to the availability of wood from standing and fallen trees, woodpeckers are frequently found. The pileated woodpecker is a typical large species. Fallen wood, known as coarse woody debris, provides shelter for many kinds of amphibians, particularly salamanders. Many well-known animals live in this kind of forest; a few examples include squirrels, which are an important canopy species, and bears, which hibernate in the winter. The top predators in deciduous forest were once wolves and cougars, along with species of weasel like the fisher.

Human Effects

Humans have often colonized areas in the temperate deciduous forest. They have harvested wood for timber and charcoal. During the settlement of North America, potash made from tree ashes were exported back to Europe as fertilizer. As a result, less than one quarter of original forests remain. Many forests are now small fragments dissected by fields and roads; these islands of green often differ substantially from the original forests, particularly along the edges. The introduction of exotic diseases continues to be a threat to forest trees, and hence, the forest; examples include the loss of chestnut and elm. At the same time, species like deer, which are clearing rather than true forest animals, have expanded their range and proliferated in these altered landscapes. Large deer populations have deleterious effects on tree regeneration overall, but particularly for edible species including yew, yellow birch and hemlock. Deer grazing also has significant negative effects on the number and kind of herbaceous flowering plants. The continuing pressure to increase deer populations, and the continued killing of top carnivores, suggests that overgrazing by deer will continue to be a significant forest conservation problem. Objective criteria for the restoration of deciduous forest include large trees, coarse woody debris, spring ephemeral, and top predators.

Freshwater Swamp Forest

Flooded forest

Freshwater swamp forests, or flooded forests, are forests which are inundated with freshwater, either permanently or seasonally. They normally occur along the lower reaches of rivers and around freshwater lakes. Freshwater swamp forests are found in a range of climate zones, from boreal through temperate and subtropical to tropical.

In the Amazon Basin of Brazil, a seasonally flooded forest is known as a *várzea*, a use that now is becoming more widespread for this type of forest in the Amazon (though generally spelled *varzea* when used in English). *Igapó*, another word used in Brazil for flooded Amazonian forests, is also sometimes used in English. Specifically, varzea refers to whitewater-inundated forest, and igapo to blackwater-inundated forest.

Peat swamp forests are swamp forests where waterlogged soils prevent woody debris from fully decomposing, which over time creates a thick layer of acidic peat.

Freshwater Swamp Forest Ecoregions

Afrotropic

- Eastern Congolian swamp forests (Democratic Republic of the Congo)
- Niger Delta swamp forests (Nigeria)
- Western Congolian swamp forests (Republic of the Congo, Democratic Republic of the Congo).

Australasia

- Northern New Guinea lowland rain and freshwater swamp forests (Indonesia, Papua New Guinea)
- Southern New Guinea freshwater swamp forests (Indonesia, Papua New Guinea)

South Asia

Ratargul Swamp Forest in Gowainghat, Sylhet, Bangladesh

- Borneo peat swamp forests (Brunei, Indonesia, Malaysia)
- Chao Phraya freshwater swamp forests (Thailand)
- Irrawaddy freshwater swamp forests (Myanmar)
- Peninsular Malaysian peat swamp forests (Malaysia, Thailand)
- Ratargul Swamp Forest (Bangladesh)
- Red River freshwater swamp forests]] (Vietnam)
- Southwest Borneo freshwater swamp forests (Indonesia)
- Tonle Sap-Mekong peat swamp forests (Cambodia, Vietnam)
- Wathurana freshwater swamp forest (Sri Lanka)
- Myristica swamp (India)
- Nelapattu Bird Sanctuary (India)

Neotropic

- Cantão igapó forest (Brazil)
- Gurupa varzea (Brazil)
- Iquitos varzea (Bolivia, Brazil, Peru)
- Marajó varzea (Brazil)
- Monte Alegre varzea (Brazil)
- Orinoco Delta swamp forests (Guyana, Venezuela)
- Pantanos de Centla (Mexico)
- Paramaribo swamp forests (Guyana, Suriname)
- Purus varzea (Brazil)

Tropical and Subtropical Moist Broadleaf Forests

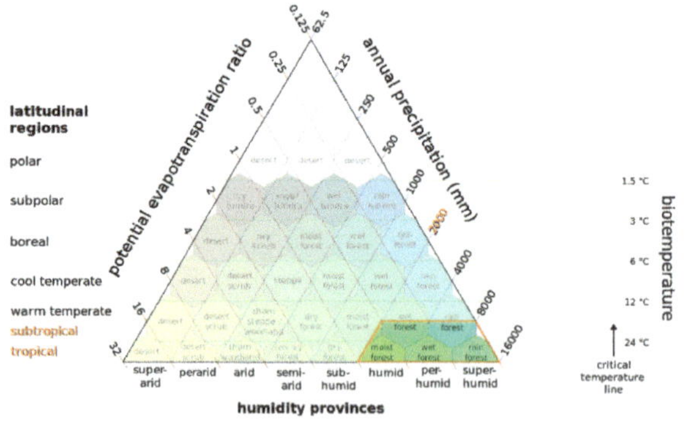

Tropical and subtropical moist forests (TSMF) as shown within the Holdridge Life Zones classification scheme, and includes moist forests, wet forests, and rainforests.

Tropical and subtropical moist broadleaf forests (TSMF), also known as tropical moist forests, are a tropical and subtropical forest biome, sometimes referred to as jungle.

Definition

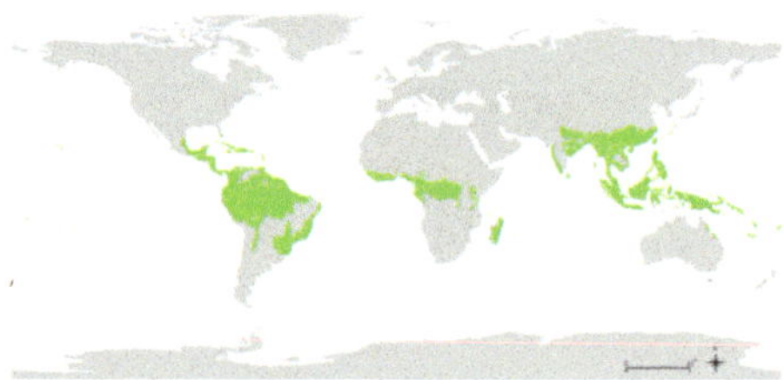

General distribution of tropical moist forests

The biome includes several types of forests:

- Lowland equatorial evergreen rain forests, commonly known as tropical rainforests, are forests which receive high rainfall (tropical rainforest climate with more than 2000 mm, or 80 inches, annually) throughout the year. These forests occur in a belt around the equator, with the largest areas in the Amazon basin of South America, the Congo basin of central Africa, and parts of the Malay Archipelago.

Rainforest lining a river bank, Cameroon

- Tropical seasonal forest: also known as moist deciduous, monsoon or semi-evergreen (mixed) seasonal forests have a monsoon or wet savannah climates (as in the Köppen climate classification): receiving high overall rainfall with a warm summer wet season and (often) a cooler winter dry season. Some trees in these forests drop some or all of their leaves during the winter dry season. These forests are found in parts of South America, in Central America and around the Caribbean, in coastal West Africa, parts of the Indian subcontinent, and across much of Indochina.
- Montane rain forests, are found in cooler-climate mountainous areas. Those with elevations high enough to regularly encounter low-level cloud cover are known as cloud forests.
- Flooded forests, including freshwater swamp forests and peat swamp forests.

In contrast to TSMF, tropical forest regions with lower levels of rainfall are home to tropical dry broadleaf forests and tropical coniferous forests. Temperate rain forests also occur in certain humid temperate coastal regions.

Occurrence

Tropical and subtropical moist broadleaf forests are common in several terrestrial ecozones, including parts of:

- The Afrotropic (equatorial Africa)
- Indomalaya (parts of the Indian subcontinent and Southeast Asia)
- The Neotropic (northern South America and Central America)
- Australasia (eastern Indonesia, New Guinea, northern and eastern Australia)
- Oceania (the tropical islands of the Pacific Ocean)

About half of the world's tropical rainforests are in the South American countries of Brazil and Peru. Rainforests now cover less than 6% of Earth's land surface. Scientists estimate that more than half of all the world's plant and animal species live in tropical rain forests.

Tropical and Subtropical Dry Broadleaf Forests

The tropical and subtropical dry broadleaf forest biome, also known as tropical dry forest, vine thicket, and vine scrub is located at tropical and subtropical latitudes. Though these forests occur in climates that are warm year-round, and may receive several hundred centimeters of rain per year, they have long dry seasons which last several months and vary with geographic location. These seasonal droughts have great impact on all living things in the forest.

Trinidad and Tobago dry forest on Chacachacare showing the dry-season deciduous nature of the vegetation

Subtropical semi-evergreen seasonal forest in Doi Inthanon National Park, Northern Thailand, at the end of the dry season.

Deciduous trees predominate in most of these forests, and during the drought a leafless period occurs, which varies with species type. Because trees lose moisture through their leaves, the shedding of leaves allows trees such as teak and mountain ebony to conserve water during dry periods. The newly bare trees open up the canopy layer, enabling sunlight to reach ground level and facilitate the growth of thick underbrush. Trees on moister sites and those with access to ground water tend to be evergreen. Infertile sites also tend to support evergreen trees. Three tropical dry broadleaf forest ecoregions, the East Deccan dry evergreen forests, the Sri Lanka dry-zone dry evergreen forests, and the Southeastern Indochina dry evergreen forests, are characterized by evergreen trees.

Though less biologically diverse than rainforests, tropical dry forests are home to a wide variety of wildlife including monkeys, deer, large cats, parrots, various rodents, and ground dwelling birds. Mammalian biomass tends to be higher in dry forests than in rain forests, especially in Asian and African dry forests. Many of these species display extraordinary adaptations to the difficult climate.

This biome is alternately known as the tropical bane forest biome or the tropical and subtropical deciduous forest biome. Locally some of these forests are also called monsoon forests, and they tend to merge into savannas.

Geographical Variation

As with seasonal tropical forests, dry forests tend to exist in the drier areas north and south of the tropical rainforest belt, south or north of the subtropical deserts, generally in two bands: one between 10° and 20°N latitude and the other between 10° and 20°S latitude. The most diverse dry forests in the world occur in southern Mexico and in the Bolivian lowlands. The dry forests of the Pacific Coast of northwestern South America support a wealth of unique species due to their dry climate. The Maputaland-Pondoland bushland and thickets along the east coast of South Africa are diverse and support many endemic species. The dry forests of central India and Indochina are notable for their diverse large vertebrate faunas. Madagascar dry deciduous forests and New Caledonia dry forests are also highly distinctive (pronounced endemism and a large number of relictual taxa) for a wide range of taxa and at higher taxonomic levels. Trees use underground water during the dry seasons.

Afrotropic

- Madagascar dry deciduous forests
    - AT0202 Madagascar dry deciduous forests

Australasia

- Nusa Tenggara Dry Forests (Indonesia)
    - AA0201 Lesser Sundas deciduous forests
    - AA0203 Sumba deciduous forests
    - AA0204 Timor and Wetar deciduous forests
- New Caledonia dry forests
    - AA0202 New Caledonia dry forests

Indomalaya

- Indochina dry forests
    - IM0202 Central Indochina dry forests
- Chhota - Nagpur dry forests
    - IM0203 Chota Nagpur dry deciduous forests

Neotropic

- Mexican dry forests
  - NT0201 Apure-Villavicencio dry forests
  - NT0204 Bajio dry forests
  - NT0205 Balsas dry forests
  - NT0227 Sierra de la Laguna dry forests
- Tumbesian - Andean valleys dry forests (Colombia, Ecuador, Peru)
  - NT0214 Ecuadorian dry forests
  - NT0221 Magdalena Valley dry forests
  - NT0223 Marañón dry forests
  - NT0232 Tumbes-Piura dry forests
- Chiquitano dry forests
  - NT0212 Chiquitano dry forests
- Atlantic dry forests
  - NT0202 Atlantic dry forests

Oceania

- Hawaii dry forests
  - OC0202 Hawaiian tropical dry forests

Biodiversity Patterns and Requirements

Species tend to have wider ranges than moist forest species, although in some regions many species do display highly restricted ranges; most dry forest species are restricted to tropical dry forests, particularly in plants; beta diversity and alpha diversity high but typically lower than adjacent moist forests.

Effective conservation of dry broadleaf forests requires the preservation of large and continuous areas of forest. Large natural areas are required to maintain larger predators and other vertebrates, and to buffer sensitive species from hunting pressure. The persistence of riparian forests and water sources is critical for many dry forest species. Large swathes of intact forest are required to allow species to recover from occasional large events, like forest fires.

Dry forests are highly sensitive to excessive burning and deforestation; overgrazing and exotic species can also quickly alter natural communities; restoration is possible but challenging, particularly if degradation has been intense and persistent.

Taiga

Taiga also boreal forest or snow forest, is a biome characterized by coniferous forests consisting mostly of pines, spruces and larches.

The taiga is the world's largest biome apart from the oceans. In North America it covers most of inland Canada and Alaska as well as parts of the extreme northern continental United States (northern Minnesota through the Upper Peninsula of Michigan to Upstate New York and northern New England), where it is known as the Northwoods. In Eurasia, it covers most of Sweden, Finland, much of Norway, some lowland/coastal areas of Iceland, much of Russia from Karelia in the west to the Pacific Ocean (including much of Siberia), and areas of northern Kazakhstan, northern Mongolia, and northern Japan (on the island of Hokkaidō). However, the main tree species, the length of the growing season and summer temperatures vary. For example, the taiga of North America mostly consists of spruces; Scandinavian and Finnish taiga consists of a mix of spruce, pines and birch; Russian taiga has spruces, pines and larches depending on the region, while the Eastern Siberian taiga is a vast larch forest.

A different use of the term taiga is often encountered in the English language, with "boreal forest" used in the United States and Canada to refer to only the more southerly part of the biome, while "taiga" is used to describe the more barren areas of the northernmost part of the biome approaching the tree line and the tundra biome. Hoffman (1958) discusses the origin of this differential use in North America and why it is an inappropriate differentiation of the Russian term. Although at high elevations taiga grades into alpine tundra through Krummholz, it is not an alpine biome only like subalpine forest, and much of taiga is lowlands.

White spruce taiga, Denali Highway, Alaska Range, Alaska

Climate and Geography

Taiga is the world's largest land biome, and makes up 29% of the world's forest cover; the largest areas are located in Russia and Canada. The taiga is the terrestrial biome with the lowest annual average temperatures after the tundra and permanent ice caps. Extreme winter minimums in the northern taiga are typically lower than those of the tundra. The lowest reliably recorded temperatures in the Northern Hemisphere were recorded in the taiga of northeastern Russia. The taiga or boreal forest has a subarctic climate with very large temperature range between seasons, but the long and cold winter is the dominant feature. This climate is classified as *Dfc*, *Dwc*, *Dsc*, *Dfd* and *Dwd* in the Köppen climate classification scheme, meaning that the short summer (24-hr average 10 °C (50 °F) or more) lasts 1–3 months and always less than 4 months. In Siberian taiga the average temperature of the coldest month is between −6 °C (21 °F) and −50 °C (−58 °F). There are also some much smaller areas grading towards the oceanic *Cfc* climate with milder winters, whilst the extreme south and (in Eurasia) west of the taiga reaches into humid continental climates (*Dfb*, *Dwb*) with longer summers. The mean annual temperature generally varies from -5 °C to 5 °C, (23 °F to 41 °F) but there are taiga areas in eastern Siberia and interior Alaska-Yukon where the mean annual reaches down to -10 °C. (14 °F) According to some sources, the boreal forest grades into a temperate mixed forest when mean annual temperature reaches about 3 °C (37 °F). Discontinuous permafrost is found in areas with mean annual temperature below 0 °C, whilst in the *Dfd* and *Dwd* climate zones continuous permafrost occurs and restricts growth to very shallow-rooted trees like Siberian larch. The winters, with average temperatures below freezing, last five to seven months. Temperatures vary from −54 °C to 30 °C (-65 °F to 86 °F) throughout the whole year. The summers, while short, are generally warm and humid. In much of the taiga, -20 °C (-4 °F) would be a typical winter day temperature and 18 °C (64 °F) an average summer day.

The taiga in the river valley near Verkhoyansk, Russia, at 67°N, must deal with the coldest winter temperatures in the northern hemisphere, but the extreme continentality of the climate gives an average daily high of 22 °C (72 °F) in July.

Boreal forest near Shovel Point in Tettegouche State Park, along the northern shore of Lake Superior in Minnesota.

The growing season, when the vegetation in the taiga comes alive, is usually slightly longer than the climatic definition of summer as the plants of the boreal biome have a lower threshold to trigger growth. In Canada, Scandinavia and Finland, the growing season is often estimated by using the period of the year when the 24-hour average temperature is +5 °C (41 °F) or more. For the Taiga Plains in Canada, growing season varies from 80 to 150 days, and in the Taiga Shield from 100 to 140 days. Some sources claim 130 days growing season as typical for the taiga. Other sources mention that 50–100 frost-free days are characteristic. Data for locations in southwest Yukon gives 80–120 frost-free days. The closed canopy boreal forest in Kenozersky National Park near Plesetsk, Arkhangelsk Province, Russia, on average has 108 frost-free days. The longest growing season is found in the smaller areas with oceanic influences; in coastal areas of Scandinavia and Finland, the growing season of the closed boreal forest can be 145–180 days. The shortest growing season is found at the northern taiga–tundra ecotone, where the northern taiga forest no longer can grow and the tundra dominates the landscape when the growing season is down to 50–70 days, and the 24-hr average of the warmest month of the year usually is 10 °C (50 °F) or less. High latitudes mean that the sun does not rise far above the horizon, and less solar energy is received than further south. But the high latitude also ensures very long summer days, as the sun stays above the horizon nearly 20 hours each day, with only around 6 hours of daylight occurring in the dark winters, depending on latitude. The areas of the taiga inside the Arctic Circle have midnight sun in mid-summer and polar night in mid-winter.

The taiga experiences relatively low precipitation throughout the year (generally 200–750 mm annually, 1,000 mm in some areas), primarily as rain during the summer months, but also as fog and snow. This fog, especially predominant in low-lying areas during and after the thawing of frozen Arctic seas, means that sunshine is not abundant in the taiga even during the long summer days. As evaporation is consequently low for most of the year, precipitation exceeds evaporation, and is sufficient to sustain the dense vegetation growth. Snow may remain on the ground for as long as nine months in the northernmost extensions of the taiga ecozone.

Lakes and other water bodies are common in the taiga. The Helvetinjärvi National Park, Finland, situated in the closed canopy taiga (mid-boreal to south-boreal) with mean annual temperature of 4 °C (39 °F).

In general, taiga grows to the south of the 10 °C July isotherm, but occasionally as far north as the 9 °C (48 °F) July isotherm. Rich in spruces, Scots pines in the western Siberian plain, the taiga is dominated by larch in Eastern Siberia, before returning to its original floristic richness on the Pacific shores. Two deciduous trees mingle throughout southern Siberia: birch and populus tremula.

Late September in the fjords near Narvik, Norway. This oceanic part of the forest can see more than 1,000 mm precipitation annually and has warmer winters than the vast inland taiga

The southern limit is more variable, depending on rainfall; taiga may be replaced by forest steppe south of the 15 °C (59 °F) July isotherm where rainfall is very low, but more typically extends south to the 18 °C (64 °F) July isotherm, and locally where rainfall is higher (notably in eastern Siberia and adjacent Outer Manchuria) south to the 20 °C (68 °F) July isotherm. In these warmer areas the taiga has higher species diversity, with more warmth-loving species such as Korean pine, Jezo spruce, and Manchurian fir, and merges gradually into mixed temperate forest or, more locally (on the Pacific Ocean coasts of North America and Asia), into coniferous temperate rainforests where oak and hornbeam appear and join the conifers, birch and populus tremula.

The area currently classified as taiga in Europe and North America (except Alaska) was recently glaciated. As the glaciers receded they left depressions in the topography that have since filled with water, creating lakes and bogs (especially muskeg soil) found throughout the taiga.

Yukon, Canada. Several of the world's longest rivers go through the taiga, including Ob, Yenisei, Lena, and Mackenzie.

In Sweden the taiga is associated with the Norrland terrain.

Soils

Taiga soil tends to be young and poor in nutrients. It lacks the deep, organically enriched profile present in temperate deciduous forests. The thinness of the soil is due largely to the cold, which hinders the development of soil and the ease with which plants can use its nutrients. Fallen leaves and moss can remain on the forest floor for a long time in the cool, moist climate, which limits their organic contribution to the soil; acids from evergreen needles further leach the soil, creating spodosol, also known as podzol. Since the soil is acidic due to the falling pine needles, the forest floor has only lichens and some mosses growing on it. In clearings in the forest and in areas with more boreal deciduous trees, there are more herbs and berries growing. Diversity of soil organisms in the boreal forest is high, comparable to the tropical rainforest.

Flora

Boreal forest near Lake Baikal in Russia

Since North America and Asia used to be connected by the Bering land bridge, a number of animal and plant species (more animals than plants) were able to colonize both continents and are distributed throughout the taiga biome. Others differ regionally, typically with each genus having several distinct species, each occupying different regions of the taiga. Taigas also have some small-leaved deciduous trees like birch, alder, willow, and poplar; mostly in areas escaping the most extreme winter cold. However, the Dahurian larch tolerates the coldest winters in the Northern Hemisphere in eastern Siberia. The very southernmost parts of the taiga may have trees such as oak, maple, elm and lime scattered among the conifers, and there is usually a gradual transition into a temperate mixed forest, such as the eastern forest-boreal transition of eastern Canada. In the interior of the continents with the driest climate, the boreal forests might grade into temperate grassland.

There are two major types of taiga. The southern part is the closed canopy forest, consisting of many closely spaced trees with mossy ground cover. In clearings in the forest, shrubs and wildflowers are common, such as the fireweed. The other type is the lichen woodland or sparse taiga, with trees that are farther-spaced and lichen ground cover; the latter is common in the northernmost taiga. In the northernmost taiga the forest cover is not only more sparse, but often stunted in growth form; moreover, ice pruned asymmetric black spruce (in North America) are often seen, with diminished foliage on the windward side. In Canada, Scandinavia and Finland, the boreal forest is usually divided into three subzones: The high boreal (north boreal) or taiga zone; the middle boreal (closed forest); and the southern boreal, a closed canopy boreal forest with some scattered temperate deciduous trees among the conifers, such as maple, elm and oak. This southern boreal forest experiences the longest and warmest growing season of the biome, and in some regions (including Scandinavia, Finland and western Russia) this subzone is commonly used for agricultural purposes. The boreal forest is home to many types of berries; some are confined to the southern and middle closed boreal forest (such as wild strawberry and partridgeberry); others grow in most areas of the taiga (such as cranberry and cloudberry), and some can grow in both the taiga and the low arctic (southern part of) tundra (such as bilberry, bunchberry and lingonberry).

The forests of the taiga are largely coniferous, dominated by larch, spruce, fir and pine. The woodland mix varies according to geography and climate so for example the Eastern Canadian forests ecoregion of the higher elevations of the Laurentian Mountains and the northern Appalachian Mountains in Canada is dominated by balsam fir *Abies balsamea*, while further north the Eastern Canadian Shield taiga of northern Quebec and Labrador is notably black spruce *Picea mariana* and tamarack larch *Larix laricina*.

Evergreen species in the taiga (spruce, fir, and pine) have a number of adaptations specifically for survival in harsh taiga winters, although larch, the most cold-tolerant of all trees, is deciduous. Taiga trees tend to have shallow roots to take advantage of the thin soils, while many of them seasonally alter their biochemistry to make them more resistant to freezing, called "hardening". The narrow conical shape of northern conifers, and their downward-drooping limbs, also help them shed snow.

Because the sun is low in the horizon for most of the year, it is difficult for plants to generate energy from photosynthesis. Pine, spruce and fir do not lose their leaves seasonally and are able to photosynthesize with their older leaves in late winter and spring when light is good but temperatures are still too low for new growth to commence. The adaptation of evergreen needles limits the water lost due to transpiration and their dark green color increases their absorption of sunlight. Although precipitation is not a limiting factor, the ground freezes during the winter months and plant roots are unable to absorb water, so desiccation can be a severe problem in late winter for evergreens.

Moss (*Ptilium crista-castrensis*) cover on the floor of taiga

Although the taiga is dominated by coniferous forests, some broadleaf trees also occur, notably birch, aspen, willow, and rowan. Many smaller herbaceous plants, such as ferns and occasionally ramps grow closer to the ground. Periodic stand-replacing wildfires (with return times of between 20–200 years) clear out the tree canopies, allowing sunlight to invigorate new growth on the forest floor. For some species, wildfires are a necessary part of the life cycle in the taiga; some, e.g. jack pine have cones which only open to release their seed after a fire, dispersing their seeds onto the newly cleared ground; certain species of fungi (such as morels) are also known to do this. Grasses grow wherever they can find a patch of sun, and mosses and lichens thrive on the damp ground and on the sides of tree trunks. In comparison with other biomes, however, the taiga has low biological diversity.

Jack pine cones and morels after fire in a boreal forest.

Coniferous trees are the dominant plants of the taiga biome. A very few species in four main genera are found: the evergreen spruce, fir and pine, and the deciduous larch. In North America, one or two species of fir and one or two species of spruce are dominant. Across Scandinavia and western Russia, the Scots pine is a common component of the taiga, while taiga of the Russian Far East and Mongolia is dominated by larch.

Fauna

Brown bear, Kamchatka peninsula. Brown bears are among the largest and most widespread taiga omnivores.

The boreal forest, or taiga, supports a relatively small range of animals due to the harshness of the climate. Canada's boreal forest includes 85 species of mammals, 130 species of fish, and an estimated 32,000 species of insects. Insects play a critical role as pollinators, decomposers, and as a part of the food web. Many nesting birds rely on them for food in the summer months. The cold winters and short summers make the taiga a challenging biome for reptiles and amphibians, which depend on environmental conditions to regulate their body temperatures, and there are only a few species in the boreal forest including red-sided garter snake, common European adder, blue-spotted salamander, northern two-lined salamander, Siberian salamander, wood frog, northern leopard frog, boreal chorus frog, American toad, and Canadian toad. Most hibernate underground in winter. Fish of the taiga must be able to withstand cold water conditions and be able to adapt to life under ice-covered water. Species in the taiga include Alaska blackfish, northern pike, walleye, longnose sucker, white sucker, various species of cisco, lake whitefish, round whitefish, pygmy whitefish, Arctic lamprey, various grayling species, brook trout (including sea-run brook trout in the Hudson Bay area), chum salmon, Siberian taimen, lenok and lake chub.

The taiga is home to a number of large herbivorous mammals, such as moose and reindeer/caribou. Some areas of the more southern closed boreal forest also have populations of other deer species such as the elk (wapiti) and roe deer. The largest animal in the taiga is the wood bison, found in northern Canada, Alaska and has been newly introduced into the Russian far-east. Small mammals of the Taiga biome include rodent species including beaver, squirrel, North American porcupine and vole, as well as a small number of lagomorph species such as snowshoe hare and mountain hare. These species have adapted to survive the harsh winters in their native ranges. Some larger mammals, such as bears, eat heartily during the summer in order to gain weight, and then go into hibernation during the winter. Other animals have adapted layers of fur or feathers to insulate them from the cold. Predatory mammals of the taiga must be adapted to travel long distances in search of scattered prey or be able to supplement their diet with vegetation or other forms of food (such as raccoons). Mammalian predators of the taiga include Canada lynx, Eurasian lynx, stoat, Siberian weasel, least weasel, sable, American marten, North American river otter, European otter, American mink, wolverine, Asian badger, fisher, gray wolf, coyote, red fox, brown bear, American black bear, Asiatic black bear, polar bear (only small areas at the taiga - tundra ecotone) and Siberian tiger.

More than 300 species of birds have their nesting grounds in the taiga. Siberian thrush, white-throated sparrow, and black-throated green warbler migrate to this habitat to take advantage of the long summer days and abundance of insects found around the numerous bogs and lakes. Of the 300 species of birds that summer in the taiga only 30 stay for the winter. These are either carrion-feeding or large raptors that can take live mammal prey, including golden eagle, rough-legged buzzard (also known as the rough-legged hawk), and raven, or else seed-eating birds, including several species of grouse and crossbills.

Fire

Fire has been one of the most important factors shaping the composition and development of boreal forest stands (Rowe 1955); it is the dominant stand-renewing disturbance through much of the Canadian boreal forest (Amiro et al. 2001). The fire history that characterizes an ecosystem is its *fire regime*, which has 3 elements: (1) fire type and intensity (e.g., crown fires, severe surface fires, and light surface fires), (2) size of typical fires of significance, and (3) frequency or return intervals for specific land units (Heinselman 1981). The average time within a fire regime to burn an area equivalent to the total area of an ecosystem is its *fire rotation* (Heinselman 1973) or *fire cycle* (Van Wagner 1978). However, as Heinselman (1981) noted, each physiographic site tends to have its own return interval, so that some areas are skipped for long periods, while others might burn two-times or more often during a nominal fire rotation.

The dominant fire regime in the boreal forest is high-intensity crown fires or severe surface fires of very large size, often more than 10,000 ha, and sometimes more than 400,000 ha (Heinselman 1981). Such fires kill entire stands. Fire rotations in the drier regions of western Canada and Alaska average 50–100 years, shorter than in the moister climates of eastern Canada, where they may average 200 years or more. Fire cycles also tend to be long near the tree line in the subarctic spruce-lichen woodlands. The longest cycles, possibly 300 years, probably occur in the western boreal in floodplain white spruce (Heinselman 1981).

Amiro et al. (2001) calculated the mean fire cycle for the period 1980 to 1999 in the

Canadian boreal forest (including taiga) at 126 years. Increased fire activity has been predicted for western Canada, but parts of eastern Canada may experience less fire in future because of greater precipitation in a warmer climate (Flannigan et al. 1998).

The mature boreal forest pattern in the south shows balsam fir dominant on well-drained sites in eastern Canada changing centrally and westward to a prominence of white spruce, with black spruce and tamarack forming the forests on peats, and with jack pine usually present on dry sites except in the extreme east, where it is absent (Rowe and Scotter 1973). The effects of fires are inextricably woven into the patterns of vegetation on the landscape, which in the east favour black spruce, paper birch, and jack pine over balsam fir, and in the west give the advantage to aspen, jack pine, black spruce, and birch over white spruce. Many investigators have reported the ubiquity of charcoal under the forest floor and in the upper soil profile, e.g., La Roi (1967). Charcoal in soils provided Bryson et al. (1965) with clues about the forest history of an area 280 km north of the then current tree line at Ennadai Lake, District Keewatin, Northwest Territories.

Two lines of evidence support the thesis that fire has always been an integral factor in the boreal forest: (1) direct, eye-witness accounts and forest-fire statistics, and (2) indirect, circumstantial evidence based on the effects of fire, as well as on persisting indicators (Rowe and Scotter 1973). The patchwork mosaic of forest stands in the boreal forest, typically with abrupt, irregular boundaries circumscribing homogenous stands, is indirect but compelling testimony to the role of fire in shaping the forest. The fact is that most boreal forest stands are less than 100 years old, and only in the rather few areas that have escaped burning are there stands of white spruce older than 250 years (Rowe and Scotter 1973). The prevalence of fire-adaptive morphologic and reproductive characteristics of many boreal plant species is further evidence pointing to a long and intimate association with fire. Seven of the ten commonest trees in the boreal forest—jack pine, lodgepole pine, aspen, balsam poplar (*Populus balsamifera*), paper birch, tamarack, black spruce—can be classed as pioneers in their adaptations for rapid invasion of open areas. White spruce shows some pioneering abilities, too, but is less able than black spruce and the pines to disperse seed at all seasons. Only balsam fir and alpine fir seem to be poorly adapted to reproduce after fire, as their cones disintegrate at maturity, leaving no seed in the crowns.

The oldest forests in the northwest boreal region, some older than 300 years, are of white spruce occurring as pure stands on moist floodplains (Rowe 1970). Here, the frequency of fire is much less than on adjacent uplands dominated by pine, black spruce and aspen. In contrast, in the Cordilleran region, fire is most frequent in the valley bottoms, decreasing upward, as shown by a mosaic of young pioneer pine and broadleaf stands below, and older spruce–fir on the slopes above (Rowe and Scotter 1973). Without fire, the boreal forest would become more and more homogeneous, with the long-lived white spruce gradually replacing pine, aspen, balsam poplar, and birch, and perhaps even black spruce, except on the peatlands (Raup and Denny 1950).

Threats

Human Activities

Plesetsk Cosmodrome is situated in the taiga

Large areas of Siberia's taiga have been harvested for lumber since the collapse of the Soviet Union. Previously, the forest was protected by the restrictions of the Soviet Forest Ministry, but with the collapse of the Union, the restrictions regarding trade with Western nations have vanished. Trees are easy to harvest and sell well, so loggers have begun harvesting Russian taiga evergreen trees for sale to nations previously forbidden by Soviet law.

In Canada, eight percent of the taiga is protected from development, the provincial government allows forest management to occur on Crown land under rigorous constraints.

The main forestry practice in the boreal forest of Canada is clearcutting, which involves cutting down most of the trees in a given area, then replanting the forest as a monocrop (one species of tree) the following season.

Some of the products from logged boreal forests include toilet paper, copy paper, newsprint, and lumber. More than 90% of boreal forest products from Canada are exported for consumption and processing in the United States. However, with the recession and fewer US homes being built, that has changed.

Some of the larger cities situated in this biome are Murmansk, Arkhangelsk, Yakutsk, Anchorage, Yellowknife, Tromsø, Luleå, and Oulu.

Most companies that harvest in Canadian forests are certified by an independent third party agency such as the Forest Stewardship Council (FSC), Sustainable Forests Initiative (SFI), or the Canadian Standards Association (CSA). While the certification process differs between these groups, all of them include forest stewardship, respect for aboriginal peoples, compliance with local, provincial or national environmental laws, forest worker safety, education and training, and other environmental, business, and social requirements. The prompt renewal of all harvest sites by planting or natural renewal is also required.

Climate Change

Seney National Wildlife Refuge

The zone of latitude occupied by the boreal forest has experienced some of the greatest temperature increases on Earth, especially during the last quarter of the twentieth century. Winter temperatures have increased more than summer temperatures. The number of days with extremely cold temperatures (e.g., −20 to −40 °C (-4 to -40 °F) has decreased irregularly but systematically in nearly all the boreal region, allowing better survival for tree-damaging insects. In summer, the daily low temperature has increased more than the daily high temperature. In Fairbanks, Alaska, the length of the frost-free season has increased from 60–90 days in the early twentieth century to about 120 days a century later. Summer warming has been shown to increase water stress and reduce tree growth in dry areas of the southern boreal forest in central Alaska, western Canada and portions of far eastern Russia. Precipitation is relatively abundant in Scandinavia, Finland, northwest Russia and eastern Canada, where a longer growth season (i.e. the period when sap flow is not impeded by frozen water) accelerate tree growth. As a consequence of this warming trend, the warmer parts of the boreal forests are susceptible to replacement by grassland, parkland or temperate forest.

In Siberia, the taiga is converting from predominantly needle-shedding larch trees to evergreen conifers in response to a warming climate. This is likely to further accelerate warming, as the evergreen trees will absorb more of the sun's rays. Given the vast size of the area, such a change has the potential to affect areas well outside of the region. In much of the boreal forest in Alaska, the growth of white spruce trees are stunted by unusually warm summers, while trees on some of the coldest fringes of the forest are experiencing faster growth than previously.

Lack of moisture in the warmer summers are also stressing the birch trees of central Alaska.

Insects

Recent years have seen outbreaks of insect pests in forest-destroying plagues: the spruce-bark beetle (*Dendroctonus rufipennis*) in Yukon and Alaska; the mountain pine beetle in British Columbia; the aspen-leaf miner; the larch sawfly; the spruce budworm (*Choristoneura fumiferana*); the spruce coneworm.

Pollution

The effect of sulphur dioxide on woody boreal forest species was investigated by Addison et al. (1984), who exposed plants growing on native soils and tailings to 15.2 μmol/m³ (0.34 ppm) of SO_2 on CO_2 assimilation rate (NAR). The Canadian maximum acceptable limit for atmospheric SO_2 is 0.34 ppm. Fumigation with SO_2 significantly reduced NAR in all species and produced visible symptoms of injury in 2–20 days. The decrease in NAR of deciduous species (trembling aspen [*Populus tremuloides*], willow [*Salix*], green alder [*Alnus viridis*], and white birch [*Betula papyrifera*]) was significantly more rapid than of conifers (white spruce, black spruce [*Picea mariana*], and jack pine [*Pinus banksiana*]) or an evergreen angiosperm (Labrador tea) growing on a fertilized Brunisol. These metabolic and visible injury responses seemed to be related to the differences in S uptake owing in part to higher gas exchange rates for deciduous species than for conifers. Conifers growing in oil sands tailings responded to SO_2 with a significantly more rapid decrease in NAR compared with those growing in the Brunisol, perhaps because of predisposing toxic material in the tailings. However, sulphur uptake and visible symptom development did not differ between conifers growing on the 2 substrates.

Acidification of precipitation by anthropogenic, acid-forming emissions has been associated with damage to vegetation and reduced forest productivity, but 2-year-old white spruce that were subjected to simulated acid rain (at pH 4.6, 3.6, and 2.6) applied weekly for 7 weeks incurred no statistically significant (P 0.05) reduction in growth during the experiment compared with the background control (pH 5.6) (Abouguendia and Baschak 1987). However, symptoms of injury were observed in all treatments, the number of plants and the number of needles affected increased with increasing rain acidity and with time. Scherbatskoy and Klein (1983) found no significant effect of chlorophyll concentration in white spruce at pH 4.3 and 2.8, but Abouguendia and Baschak (1987) found a significant reduction in white spruce at pH 2.6, while the foliar sulphur content significantly greater at pH 2.6 than any of the other treatments.

Protection

Many nations are taking direct steps to protect the ecology of the taiga by prohibiting logging, mining, oil and gas production, and other forms of development. In February 2010 the Canadian government established protection for 13,000 square kilometres of boreal forest by creating a new 10,700-square-kilometre park reserve in the Mealy Mountains area of eastern Canada and a 3,000-square-kilometre waterway provincial park that follows alongside the Eagle River from headwaters to sea.

Peat bog in Dalarna, Sweden. Bogs and peatland are widespread in the taiga. They are home to a unique flora, and store vast amounts of carbon. In western Eurasia, the Scots pine is common in the boreal forest.

Two Canadian provincial governments, Ontario and Quebec, introduced measures in 2008 that would protect at least half of their northern boreal forest. Although both provinces admitted it will take years to plan, work with Aboriginal and local communities and ultimately map out precise boundaries of the areas off-limits to development, the measures are expected to create some of the largest protected areas networks in the world once completed. Both announcements came the following year after a letter signed by 1,500 scientists called on political leaders to protect at least half of the boreal forest.

The taiga stores enormous quantities of carbon, more than the world's temperate and tropical forests combined, much of it in wetlands and peatland. In fact, current estimates place boreal forests as storing twice as much carbon per unit area as tropical forests.

Natural Disturbance

One of the biggest areas of research and a topic still full of unsolved questions is the recurring disturbance of fire and the role it plays in propagating the lichen woodland. The phenomenon of wildfire by lightning strike is the primary determinant of understory vegetation and because of this, it is considered to be the predominant force behind community and ecosystem properties in the lichen woodland. The significance of fire is clearly evident when one considers that understory vegetation influences tree seedling germination in the short term and decomposition of biomass and nutrient availability in the long term. The recurrent cycle of large, damaging fire occurs approximately every 70 to 100 years. Understanding the dynamics of this ecosystem is entangled with discovering the successional paths that the vegetation exhibits after a fire. Trees, shrubs, and lichens all recover from fire-induced damage through vegetative reproduction as well as invasion by propagules. Seeds that have fallen and become buried provide little help in re-establishment of a species. The reappearance of lichens is reasoned to occur because of varying conditions and light/nutrient availability in each different microstate. Several different studies have been done that have led to the formation of the theory that post-fire development can be propagated by any of four pathways: self replacement, species-dominance relay, species replacement, or gap-phase self replacement. Self replacement is simply the re-establishment of the pre-fire dominant species. Species-dominance relay is a sequential attempt of tree species to establish dominance in the canopy. Species replacement is when fires occur in sufficient frequency to interrupt species dominance relay. Gap-Phase Self-Replacement is the least common and so far has only been documented in Western Canada. It is a self replacement of the surviving species into the canopy gaps after a fire kills another species. The particular pathway taken after a fire disturbance depends on how the landscape is able to support trees as well as fire frequency. Fire frequency has a large role in shaping the original inception of the lower forest line of the lichen woodland taiga.

It has been hypothesized by Serge Payette that the spruce-moss forest ecosystem was changed into the lichen woodland biome due to the initiation of two compounded strong disturbances: large fire and the appearance and attack of the spruce budworm. The spruce budworm is a deadly insect to the spruce populations in the southern regions of the taiga. J.P. Jasinski confirmed this theory five years later stating "Their [lichen woodlands] persistence, along with their previous moss forest histories and current occurrence adjacent to closed moss forests, indicate that they are an alternative stable state to the spruce–moss forests"

References

- Bourgeron, Patrick S. (1983). "Spatial Aspects of Vegetation Structure". In Frank B. Golley. Tropical Rain Forest Ecosystems. Structure and Function. Ecosystems of the World (14A ed.). Elsevier Scientific. pp. 29–47. ISBN 0-444-41986-1.
- Воробьев Г.И., ed. (1985). "Бор". Лесная энциклопедия (in Russian). Советская энциклопедия. Retrieved 1 March 2015.
- "Study reveals for first time true diversity of life in soils across the globe, new species discovered". Physorg.com. Retrieved 2012-01-14.
- "Russian boreal forests undergoing vegetation change, study shows". Sciencedaily.com. 2011-03-25. doi:10.1111/j.1365-2486.2011.02417.x. Retrieved 2012-01-14.
- "Fairbanks Daily News-Miner - New study states boreal forests shifting as Alaska warms". News-miner.com. Retrieved 2012-01-14.
- "Final Paper: The Medicinal Value of the Rainforest May 15, 2003. Amanda Haidet May 2003". Jrscience.wcp.muohio.edu. Retrieved 2012-08-26.
- "Science: Satellite monitors Madagascar's shrinking rainforest, 19 May 1990, New Scientist". Newscientist.com. 1990-05-19. Retrieved 2012-08-26.
- "Spruce budworm and sustainable management of the boreal forest". Cfs.nrcan.gc.ca. 2007-12-05. Retrieved 2011-02-21.

- "TAMPERE/PIRKKALA, FINLAND Weather History and Climate Data". Worldclimate.com. 2007-02-04. Retrieved 2011-02-21.
- Braun, David (February 7, 2010). "Boreal landscapes added to Canada's parks Boreal landscapes added to Canada's parks". NatGeo News Watch: News Editor David Braun's Eye on the World. National Geographic Society. Retrieved 17 February 2010.
- Benjamin, Alison (6 December 2007). "More than half of Amazon will be lost by 2030, report warns". guardian.co.uk. London. Retrieved 7 July 2010.

3

Forest Management: An Integrated Study

Forest management manages the administration, economy, the social aspects and technical aspects of forestry. The restoration of forestry is defined as ecoforestry. Ecoforestry takes a holistic approach in preserving forests and the organisms in it. The aspects elucidated in this chapter are of vital importance, and provide a better understanding of forest management.

Forest Management

Forest management is a branch of forestry concerned with overall administrative, economic, legal, and social aspects, as well as scientific and technical aspects, such as silviculture, protection, and forest regulation. This includes management for aesthetics, fish, recreation, urban values, water, wilderness, wildlife, wood products, forest genetic resources, and other forest resource values. Management can be based on conservation, economics, or a mixture of the two. Techniques include timber extraction, planting and replanting of various species, cutting roads and pathways through forests, and preventing fire.

Definition

The forest is a natural system that can supply different products and services. The working of this system is influenced by the natural environment: climate, topography, soil, etc., and also by human activity. The actions of humans in forests constitute forest management. In developed societies, this management tends to be elaborate and planned in order to achieve the objectives that are considered desirable.

Some forests have been and are managed to obtain traditional forest products such as firewood, fiber for paper, and timber, with little thinking for other products and services. Nevertheless, as a result of the progression of environmental awareness, management of forests for multiple use is becoming more common.

Public Input and Awareness

There has been increased public awareness of natural resource policy, including forest management. Public concern regarding forest management may have shifted from the extraction of timber for earning money for the economy, to the preservation of additional forest resources, including wildlife and old growth forest, protecting biodiversity, watershed management, and recreation. Increased environmental awareness may contribute to an increased public mistrust of forest management professionals. But it can also lead to greater understanding about what professionals do re forests for nature conservation and ecological services. The importance of taking care of the forests for ecological as well as economical sustainable reasons has been shown in the TV show Ax Men.

The ecolabel of the Forest Stewardship Council for certified sustainable forest product.

Many tools like GIS and photogrammetry modelling have been developed to improve forest inventory and management planning.

Wildlife Considerations

The abundance and diversity of birds, mammals, amphibians and other wildlife are affected by strategies and types of forest management.

Management Intensity

Forest management varies in intensity from a leave alone, natural situation to a highly intensive regime with silvicultural interventions. Management is generally increased in intensity to achieve either economic criteria (increased timber yields, non-timber forest products, ecosystem services) or ecological criteria (species recovery, fostering of rare species, carbon sequestration).

Ecoforestry

Ecoforestry has been defined as selection forestry or restoration forestry. The main idea of Ecoforestry is to maintain or restore the forest to standards where the forest may still be harvested for products on a sustainable basis. Ecoforestry is forestry that emphasizes holistic practices which strive to protect and restore ecosystems rather than maximize economic productivity. Sustainability of the forest also comes with uncertainties. There are other factors that may affect the forest furthermore than that of the harvesting. There are internal conditions such as effects of soil compaction, tree damage, disease, fire, and blow down that also directly affect the ecosystem. These factors have to be taken into account when determining the sustainability of a forest. If these factors are added to the harvesting and production that comes out of the forest, then the forest will become less likely to survive, and will then become less sustainable.

Since the forest is considered an ecosystem, it is dependent on all of the living and non-living factors within itself. This is a major part of why the forest needs to be sustainable before it is harvested. For example, a tree would convert sunlight to sugars for respiration to keep the tree alive. The remains of the converted sugars is left in roots for consumption by the organisms surrounding the tree in the habitat. This shows the productivity of an ecosystem with its inhabitants. Productivity within the ecosystem cannot come to fruition unless the forest is sustainable enough to be harvested. If all of the organisms of the ecosystem are non existent, the ecosystem itself will soon come to be non existent. Once that happens, there is no longer any forest to harvest from. The overall productivity of a system can be found in an equation where the Net Primary Production, or NPP, is equal to the Gross Primary Production, or GPP, minus the Respiration, or R. The formula is the NPP = GPP - R. The NPP is the overall efficiency of the plants in the ecosystem. Through having a constant efficiency in NPP, the ecosystem is then more sustainable. The GPP refers to the rate of energy stored by photosynthesis in plants. The R refers to the maintenance and reproduction of plants from the energy expended.

Ecoforestry has many principles within the existence of itself. It covers sustainable development and the fair harvesting of the organisms living within the forest ecosystem. There have been many proposals of principles outlined for ecoforestry. They are covered over books, articles, and environmental agencies. All of the principles relate to the idea that in ecoforestry, less should be harvested, and diversity must be managed. Through harvesting less, there is enough biomass left in the forest, so that the forest may stay healthy and still stay maintained. It will grow at a sustainable level annually, and thus it will be able to still be harvested the following year. Through management of the diversity, species may cohabitate in an ecosystem where the forest may feed off of other species in its growth and production. The Principles of Ecoforestry may be found below.

Principles of Ecoforestry

The principles of ecoforestry are:

Number	Principle
1	One must consider retention. It must be the first consideration in any planned removal of trees from a stand. By placing retention at the beginning of a planned removal, one makes sure that they know what must be left to ensure the protection of such things as rare species, sites of native cultural significance, riparian zones.

2	Riparian zones should not be touched. It is a sensitive area, thus tree removal should not occur there. This protects the water quality. Water quality is protected by not altering the drainage pattern of the zones. No tree removal should take place in the most sensitive areas.
3	The composition and structures should be upheld so that forests may fully function. This may be subject to large old trees, snags, and large fallen trees. These pieces of the ecosystem are upheld through letting them grow and die out into a timber extraction area.
4	When removing trees, the lowest impact should be used. This means not compacting the soil in the forest and building small or no roads if at all possible. This takes away the disturbance that would be done to the foundation of the forest's grounds.
5	Plan in terms of the needs of the larger watershed. The watershed zone plan designates areas where tree removal is not permitted. It also designates the areas where removal is possible, and the different types of removal available.
6	Prohibit clear cutting. Clear cutting is not ecologically correct. There are methods for cutting trees to maintain structure, and to allow the trees to cohabitate in a healthy ecosystem.
7	Select trees as candidates for removal by considering how abundant and redundant their structures and functions are to the rest of the forest as a whole, leaving potential wildlife trees (to become snags and coarse woody debris).
8	The forest should be allowed to regenerate trees through the seeds from the trees in the logged areas. This allows tree planting to be taken out of the ecoforestry principles, allowing a natural regeneration.
9	Ecological succession should be kept at all times. This will protect biological diversity. This even means eliminating the process of brush control.
10	No slash burning. Fire may be used as a tool in landscapes that have a history of naturally occurring fires. Although this may be true, fire should still be used with caution.
11	No pesticide use. The forest needs disease, insects, and shrub/herb vegetation. They are essential parts of a fully functioning forest, even though they involve the forest's decay in some cases. They are natural parts that allow the ecosystem to operate as a whole.
12	Maintain and restore topsoil quality. This can be done through leaving sufficiently coarse and small debris on the forest grounds.
13	Maintain beauty and other natural aesthetic qualities in the visual, sound, and odor landscapes. This includes not taking away the wildlife, the plants, or structural layout of the ecosystem.
14	Always look at the forest as a whole. Each part of the forest contributes to its overall needs and health. That is how the forest would have survived without human interference.
15	Rely more on people and markets. Use accounting and budgeting as a solution to rely less on the destruction or harvesting from the forests.
16	Don't do wrong. If it feels wrong, then it is probably wrong. Don't allow ignorance to persevere. Recognize that the ecosystem needs tending to. If the forest is not preserved, then it cannot be harvested forever.

Sustainable Forest Management

Sustainable forest management (SFM) is the management of forests according to the principles of sustainable development. Sustainable forest management has to keep the balance between three main pillars: ecological, economic and socio-cultural. Successfully achieving sustainable forest management will provide integrated benefits to all, ranging from safeguarding local livelihoods to protecting the biodiversity and ecosystems provided by forests, reducing rural poverty and mitigating some of the effects of climate change.

The "Forest Principles" adopted at The United Nations Conference on Environment and Development (UNCED) in Rio de Janeiro in 1992 captured the general international understanding of sustainable forest management at that time. A number of sets of criteria and indicators have since been developed to evaluate the achievement of SFM at the global, regional, country and management unit level. These were all attempts to codify and provide for independent assessment of the degree to which the broader objectives of sustainable forest management are being achieved in practice. In 2007, the United Nations General Assembly adopted the Non-Legally Binding Instrument on All Types of Forests. The instrument was the first of its kind, and reflected the strong international commitment to promote implementation of sustainable forest management through a new approach that brings all stakeholders together.

Definition

A definition of SFM was developed by the Ministerial Conference on the Protection of Forests in Europe (FOREST EUROPE), and has since been adopted by the Food and Agriculture Organization (FAO). It defines sustainable forest management as:

The stewardship and use of forests and forest lands in a way, and at a rate, that maintains their biodiversity, productivity, regeneration capacity, vitality and their potential to fulfill, now and in the future, relevant ecological, economic and social functions, at local, national, and global levels, and that does not cause damage to other ecosystems.

In simpler terms, the concept can be described as the attainment of balance – balance between society's increasing demands for forest products and benefits, and the preservation of forest health and diversity. This balance is critical to the survival of forests, and to the prosperity of forest-dependent communities.

For forest managers, sustainably managing a particular forest tract means determining, in a tangible way, how to use it today to ensure similar benefits, health and productivity in the future. Forest managers must assess and integrate a wide array of sometimes conflicting factors – commercial and non-commercial values, environmental considerations, community needs, even global impact – to produce sound forest plans. In most cases, forest managers develop their forest plans in consultation with citizens, businesses, organizations and other interested parties in and around the forest tract being managed. The tools and visualization have been recently evolving for better management practices.

In 2014, at the request of Member States, the Food and Agriculture Organization of the United Nations developed and launched the Sustainable Forest Management Toolbox, an online collection of tools, case studies and other resources to support countries implementing sustainable forest management.

Because forests and societies are in constant flux, the desired outcome of sustainable forest management is not a fixed one. What constitutes a sustainably managed forest will change over time as values held by the public change.

Criteria and Indicators

Deforestation of native rain forest in Rio de Janeiro City for extraction of clay for civil engineering (2009 picture). An example of non sustainable forest management.

Criteria and indicators are tools which can be used to conceptualise, evaluate and implement sustainable forest management. Criteria define and characterize the essential elements, as well as a set of conditions or processes, by which sustainable forest management may be assessed. Periodically measured indicators reveal the direction of change with respect to each criterion.

Criteria and indicators of sustainable forest management are widely used and many countries produce national reports that assess their progress toward sustainable forest management. There are nine international and regional criteria and indicators initiatives, which collectively involve more than 150 countries. Three of the more advanced initiatives are those of the Working Group on Criteria and Indicators for the Conservation and Sustainable Management of Temperate and Boreal Forests (also called the Montreal Process), Forest Europe, and the International Tropical Timber Organization. Countries who are members of the same initiative usually agree to produce reports at the same time and using the same indicators. Within countries, at the management unit level, efforts have also been directed at developing local level criteria and indicators of sustainable forest management. The Center for International Forestry Research, the International Model Forest Network and researchers at the University of British Columbia have developed a number of tools and techniques to help forest-dependent communities develop their own local level criteria and indicators. Criteria and Indicators also form the basis of third-party forest certification programs such as the Canadian Standards Association's Sustainable Forest Management Standards and the Sustainable Forestry Initiative Standard.

There appears to be growing international consensus on the key elements of sustainable forest management. Seven common thematic areas of sustainable forest management have emerged based on the criteria of the nine ongoing regional and international criteria and indicators initiatives. The seven thematic areas are:

- Extent of forest resources
- Biological diversity
- Forest health and vitality
- Productive functions and forest resources
- Protective functions of forest resources
- Socio-economic functions
- Legal, policy and institutional framework.

This consensus on common thematic areas (or criteria) effectively provides a common, implicit definition of sustainable forest management. The seven thematic areas were acknowledged by the international forest community at the fourth session of the United Nations Forum on Forests and the 16th session of the Committee on Forestry. These thematic areas have since been enshrined in the Non-Legally Binding Instrument on All Types of Forests as a reference framework for sustainable forest management to help achieve the purpose of the instrument.

On January 5, 2012, the Montreal Process, Forest Europe, the International Tropical Timber Organization, and the Food and Agriculture Organization of the United Nations, acknowledging the seven thematic areas, endorsed a joint statement of collaboration to improve global forest related data collection and reporting and avoiding the proliferation of monitoring requirements and associated reporting burdens.

Ecosystem Approach

The Ecosystem Approach has been prominent on the agenda of the Convention on Biological Diversity (CBD) since 1995 . The CBD definition of the Ecosystem Approach and a set of principles for its application were developed at an expert meeting in Malawi in 1995, known as the Malawi Principles. The definition, 12 principles and 5 points of "operational guidance" were adopted by the fifth Conference of Parties (COP5) in 2000. The CBD definition is as follows

The ecosystem approach is a strategy for the integrated management of land, water and living resources that promotes conservation and sustainable use in an equitable way. Application of the ecosystem approach will help to reach a balance of the three objectives of the Convention. An ecosystem approach is based on the application of appropriate scientific methodologies focused on levels of biological organization, which encompasses the essential structures, processes, functions and interactions among organisms and their environment. It recognizes that humans, with their cultural diversity, are an integral component of many ecosystems.

Sustainable forest management was recognized by parties to the Convention on Biological Diversity in 2004 (Decision VII/11 of COP7) to be a concrete means of applying the Ecosystem Approach to forest ecosystems. The two concepts, sustainable forest management and the ecosystem approach, aim at promoting conservation and management practices which are environmentally, socially and economically sustainable, and which generate and maintain benefits for both present and future generations. In Europe, the MCPFE and the Council for the Pan-European Biological and Landscape Diversity Strategy (PEBLDS) jointly recognized sustainable forest management to be consistent with the Ecosystem Approach in 2006.

Independent Certification

Growing environmental awareness and consumer demand for more socially responsible businesses helped third-party forest certification emerge in the 1990s as a credible tool for communicating the environmental and social performance of forest operations.

There are many potential users of certification, including: forest managers, scientists, policy makers, investors, environmental advocates, business consumers of wood and paper, and individuals.

With third-party forest certification, an independent organization develops standards of good forest management, and independent auditors issue certificates to forest operations that comply with those standards. Forest certification verifies that forests are well-managed—as defined by a particular standard—and chain-of-custody certification tracks wood and paper products from the certified forest through processing to the point of sale.

This rise of certification led to the emergence of several different systems throughout the world. As a result, there is no single accepted forest management standard worldwide, and each system takes a somewhat different approach in defining standards for sustainable forest management.

In its 2009–2010 Forest Products Annual Market Review United Nations Economic Commission for Europe/Food and Agriculture Organization stated: "Over the years, many of the issues that previously divided the (certification) systems have become much less distinct. The largest certification systems now generally have the same structural programmatic requirements. "

Third-party forest certification is an important tool for those seeking to ensure that the paper and wood products they purchase and use come from forests that are well-managed and legally harvested. Incorporating third-party certification into forest product procurement practices can be a centerpiece for comprehensive wood and paper policies that include factors such as the protection of sensitive forest values, thoughtful material selection and efficient use of products.

There are more than fifty certification standards worldwide, addressing the diversity of forest types and tenures. Globally, the two largest umbrella certification programs are:

- Programme for the Endorsement of Forest Certification (PEFC)
- Forest Stewardship Council (FSC)

The area of forest certified worldwide is growing slowly. PEFC is the world's largest forest certification system, with more than two-thirds of the total global certified area certified to its Sustainability Benchmarks.

In North America, there are three certification standards endorsed by PEFC – the Sustainable Forestry Initiative, the Canadian Standards Association's Sustainable Forest Management Standard, and the American Tree Farm System. FSC has five standards in North America – one in the United States and four in Canada.

While certification is intended as a tool to enhance forest management practices throughout the world, to date most certified forestry operations are located in Europe and North America. A significant barrier for many forest managers in developing countries is that they lack the capacity to undergo a certification audit and maintain operations to a certification standard.

Forest Governance

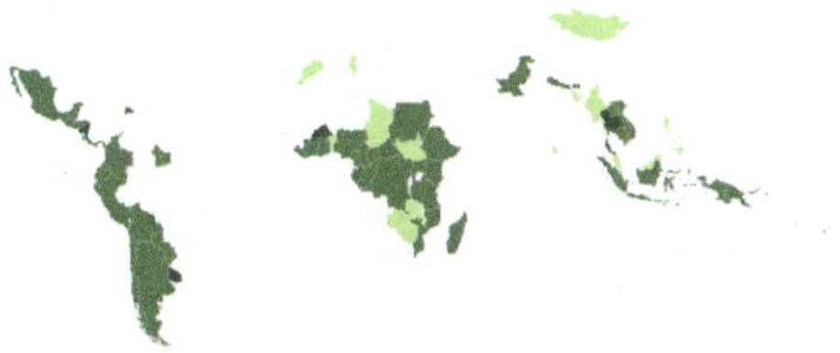

Countries participating in the UNREDD program and/or Forest Carbon Partnership Facility.

UN-REDD participants

Forest Carbon Partnership Facility participants

participants in both

Although a majority of forests continue to be owned formally by government, the effectiveness of forest governance is increasingly independent of formal ownership. Since neo-liberal ideology in the 1980s and the emanation of the climate change challenges, evidence that the state is failing to effectively manage environmental resources has emerged. Under neo-liberal regimes in the developing countries, the role of the state has diminished and the market forces have increasingly taken over the dominant socio-economic role. Though the critiques of neo-liberal policies have maintained that market forces are not only inappropriate for sustaining the environment, but are in fact a major cause of environmental destruction. Hardin's tragedy of the common (1968) has shown that the people cannot be left to do as they wish with land or environmental resources. Thus, decentralization of management offers an alternative solution to forest governance.

The shifting of natural resource management responsibilities from central to state and local governments, where this is occurring, is usually a part of broader decentralization process. According to Rondinelli and Cheema (1983), there are four distinct decentralization options: these are: (i) Privatization – the transfer of authority from the central government to non-governmental sectors otherwise known as market-based service provision, (ii) Delegation – centrally nominated local authority, (iii) Devolution – transfer of power to locally acceptable authority and (iv) Deconcentration – the redistribution of authority from the central government to field delegations of the central government. The major key to effective decentralization is increased broad-based participation in local-public decision making. In 2000, the World Bank report reveals that local government knows the needs and desires of their constituents better than the national government, while at the same time, it is easier to hold local leaders accountable. From the study of West African tropical forest, it is argued that the downwardly accountable and/or representative authorities with meaningful discretional powers are the basic institutional element of decentralization that should lead to efficiency, development and equity. This collaborates with the World Bank report in 2000 which says that decentralization should improve resource allocation, efficiency, accountability and equity "by linking the cost and benefit of local services more closely".

Many reasons point to the advocacy of decentralization of forest. (i) Integrated rural development projects often fail because they are top-down projects that did not take local people's needs and desires into account. (ii) National government sometimes have legal authority over vast forest areas that they cannot control, thus, many protected area projects result in increased biodiversity loss and greater social conflict. Within the sphere of forest management, as state earlier, the most effective option of decentralization is "devolution"-the transfer of power to locally accountable authority. However, apprehension about local governments is not unfounded. They are often short of resources, may be staffed by people with low education and are sometimes captured by local elites who promote clientelist relation rather than democratic participation. Enters and Anderson (1999) point that the result of community-based projects intended to reverse the problems of past central approaches to conservation and development have also been discouraging.

Broadly speaking, the goal of forest conservation has historically not been met when, in contrast with land use changes; driven by demand for food, fuel and profit. It is necessary to recognize and advocate for better forest governance more strongly given the importance of forest in meeting basic human needs in the future and maintaining ecosystem and biodiversity as well as addressing climate change mitigation and adaptation goal. Such advocacy must be coupled with financial incentives for government of developing countries and greater governance role for local government, civil society, private sector and NGOs on behalf of the "communities".

National Forest Funds

The development of National Forest Funds is one way to address the issue of financing sustainable forest management. National forest funds (NFFs) are dedicated financing mechanisms managed by public institutions designed to support the conservation and sustainable use of forest resources. As of 2014, there are 70 NFFs operating globally.

Plantation

A pine plantation in the United States. Tree plantations are usually easily distinguished from natural forests by the trees being planted in straight lines.

One operator planting poplar cuttings with a Rotor machine for planting a new nursery.

A plantation is a large piece of land (or water) where one crop is specifically planted for widespread commercial sale and usually tended by resident laborers. The crops grown include fast-growing trees (often conifers), cotton, coffee, tea, cocoa, sugar cane, sisal, oil seeds (e.g. oil palms), rubber trees, and various fruits. Protectionist policies and natural comparative advantage have sometimes contributed to determining where plantations were located.

Among the earliest examples of plantations were the latifundia of the Roman Empire, which produced large quantities of wine and olive oil for export. Plantation agriculture grew rapidly with the increase in international trade and the development of a worldwide economy that followed the expansion of European colonial empires. Like every economic activity, it has changed over time. Earlier forms of plantation agriculture were associated with large disparities of wealth and income, foreign ownership and political influence, and exploitative social systems such as indentured labour and slavery.

Types

Industrial

A plantation of Douglas-fir in Washington, U.S.

Industrial plantations are established to produce a high volume of wood in a short period of time. Plantations are grown by state forestry authorities (for example, the Forestry Commission in Britain) and/or the paper and wood industries and other private landowners (such as Weyerhaeuser, Rayonier and Sierra Pacific Industries in the United States, Asia Pulp & Paper in Indonesia). Christmas trees are often grown on plantations as well. In southern and southeastern Asia, teak plantations have recently replaced the natural forest.

Machine for industrial plantations

A stump grinder working to remove a tree stump.

Industrial plantations are actively managed for the commercial production of forest products. Industrial plantations are usually large-scale. Individual blocks are usually even-aged and often consist of just one or two species. These species can be exotic or indigenous. The plants used for the plantation are often genetically altered for desired traits such as growth and resistance to pests and diseases in general and specific traits, for example in the case of timber species, volumic wood production and stem straightness. Forest genetic resources are the basis for genetic alteration. Selected individuals grown in seed orchards are a good source for seeds to develop adequate planting material.

Wood production on a tree plantation is generally higher than that of natural forests. While forests managed for wood production commonly yield between 1 and 3 cubic meters per hectare per year, plantations of fast-growing species commonly yield between 20 and 30 cubic meters or more per hectare annually; a Grand Fir plantation at Craigvinean in Scotland has a growth rate of 34 cubic meters per hectare per year (Aldhous & Low 1974), and Monterey Pine plantations in southern Australia can yield up to 40 cubic meters per hectare per year (Everard & Fourt 1974). In 2000, while plantations accounted for 5% of global forest, it is estimated that they supplied about 35% of the world's roundwood.

Growth Cycle

- In the first year, the ground is prepared usually by the combination of burning, herbicide spraying, and/or cultivation and then saplings are planted by human crew or by machine. The saplings are usually obtained in bulk from industrial nurseries, which may specialize in selective breeding in order to produce fast growing disease- and pest-resistant strains.
- In the first few years until the canopy closes, the saplings are looked after, and may be dusted or sprayed with fertilizers or pesticides until established.
- After the canopy closes, with the tree crowns touching each other, the plantation is becoming dense and crowded, and tree growth is slowing due to competition. This stage is termed 'pole stage'. When competition becomes too intense (for

pine trees, when the live crown is less than a third of the tree's total height), it is time to thin out the section. There are several methods for thinning, but where topography permits, the most popular is 'row-thinning', where every third or fourth or fifth row of trees is removed, usually with a harvester. Many trees are removed, leaving regular clear lanes through the section so that the remaining trees have room to expand again. The removed trees are delimbed, forwarded to the forest road, loaded onto trucks, and sent to a mill. A typical pole stage plantation tree is 7–30 cm in diameter at breast height (dbh). Such trees are sometimes not suitable for timber, but are used as pulp for paper and particle-board, and as chips for oriented strand board.

- As the trees grow and become dense and crowded again, the thinning process is repeated. Depending on growth rate and species, trees at this age may be large enough for timber milling; if not, they are again used as pulp and chips.
- Around year 10-60 the plantation is now mature and (in economic terms) is falling off the back side of its growth curve. That is to say, it is passing the point of maximum wood growth per hectare per year, and so is ready for the final harvest. All remaining trees are felled, delimbed, and taken to be processed.
- The ground is cleared, and the cycle is repeated.

Some plantation trees, such as pines and eucalyptus, can be at high risk of fire damage because their leaf oils and resins are flammable to the point of a tree being explosive under some conditions. Conversely, an afflicted plantation can in some cases be cleared of pest species cheaply through the use of a prescribed burn, which kills all lesser plants but does not significantly harm the mature trees.

Natural Forest Loss

Many forestry experts claim that the establishment of plantations will reduce or eliminate the need to exploit natural forest for wood production. In principle this is true because due to the high productivity of plantations less land is needed. Many point to the example of New Zealand, where 19% of the forest area provides 99% of the supply of industrial round wood. It has been estimated that the world's demand for fiber could be met by just 5% of the world forest (Sedjo & Botkin 1997). However, in practice, plantations are replacing natural forest, for example in Indonesia. According to the FAO, about 7% of the natural closed forest being lost in the tropics is land being converted to plantations. The remaining 93% of the loss is land being converted to agriculture and other uses. Worldwide, an estimated 15% of plantations in tropical countries are established on closed canopy natural forest.

In the Kyoto Protocol, there are proposals encouraging the use of plantations to reduce carbon dioxide levels (though this idea is being challenged by some groups on the grounds that the sequestered CO_2 is eventually released after harvest).

A tea plantation in Ciwidey, Bandung in Indonesia

Criticism

Bushfires pose a high risk to Eucalyptus plantations.

In contrast to a naturally regenerated forest, plantations are typically grown as even-aged monocultures, primarily for timber production.

- Plantations are usually near- or total monocultures. That is, the same species of tree is planted across a given area, whereas a natural forest would contain a far more diverse range of tree species.
- Plantations may include tree species that would not naturally occur in the area. They may include unconventional types such as hybrids, and genetically modified trees may be used sometime in the future. Since the primary interest in plantations is to produce wood or pulp, the types of trees found in plantations are those that are best-suited to industrial applications. For example, pine, spruce and eucalyptus are widely planted far beyond their natural range because of their fast growth rate, tolerance of rich or degraded agricultural land and potential to produce large volumes of raw material for industrial use.
- Plantations are always young forests in ecological terms. Typically, trees grown

in plantations are harvested after 10 to 60 years, rarely up to 120 years. This means that the forests produced by plantations do not contain the type of growth, soil or wildlife typical of old-growth natural forest ecosystems. Most conspicuous is the absence of decaying dead wood, a crucial component of natural forest ecosystems.

In the 1970s, Brazil began to establish high-yield, intensively managed, short rotation plantations. These types of plantations are sometimes called fast-wood plantations or fiber farms and often managed on a short-rotation basis, as little as 5 to 15 years. They are becoming more widespread in South America, Asia and other areas. The environmental and social impacts of this type of plantation has caused them to become controversial. In Indonesia, for example, large multi-national pulp companies have harvested large areas of natural forest without regard for regeneration. From 1980 to 2000, about 50% of the 1.4 million hectares of pulpwood plantations in Indonesia have been established on what was formerly natural forest land.

The replacement of natural forest with tree plantations has also caused social problems. In some countries, again, notably Indonesia, conversions of natural forest are made with little regard for rights of the local people. Plantations established purely for the production of fiber provide a much narrower range of services than the original natural forest for the local people. India has sought to limit this damage by limiting the amount of land owned by one entity and, as a result, smaller plantations are owned by local farmers who then sell the wood to larger companies. Some large environmental organizations are critical of these high-yield plantations and are running an anti-plantation campaign, notably the Rainforest Action Network and Greenpeace.

Farm and Home

Farm or home plantations are typically established for the production of timber and fire wood for home use and sometimes for sale. Management may be less intensive than with Industrial plantations. In time, this type of plantation can become difficult to distinguish from naturally regenerated forest.

Teak and bamboo plantations in India have given good results and an alternative crop solution to farmers of central India, where conventional farming was popular. But due to rising input costs of farming many farmers have done teak and bamboo plantations which require very little water (only during first two years). Teak and bamboo have legal protection from theft. Bamboo, once planted, gives output for 50 years till flowering occurs. Teak requires 20 years to grow to full maturity and fetch returns.

Environmental

These may be established for watershed or soil protection. They are established for erosion control, landslide stabilization and windbreaks. Such plantations are established to foster native species and promote forest regeneration on degraded lands as a tool of environmental restoration.

Ecological Impact

The last batch of sawnwood from the peat forest in Indragiri Hulu, Sumatra, Indonesia. Deforestation for oil palm plantation.

Probably the single most important factor a plantation has on the local environment is the site where the plantation is established. If natural forest is cleared for a planted forest then a reduction in biodiversity and loss of habitat will likely result. In some cases, their establishment may involve draining wetlands to replace mixed hardwoods that formerly predominated with pine species. If a plantation is established on abandoned agricultural land, or highly degraded land, it can result in an increase in both habitat and biodiversity. A planted forest can be profitably established on lands that will not support agriculture or suffer from lack of natural regeneration.

The tree species used in a plantation is also an important factor. Where non-native varieties or species are grown, few of the native fauna are adapted to exploit these and further biodiversity loss occurs. However, even non-native tree species may serve as corridors for wildlife and act as a buffer for native forest, reducing edge effect.

Once a plantation is established, how it is managed becomes the important environmental factor. The single most important factor of management is the rotation period. Plantations harvested on longer rotation periods (30 years or more) can provide similar benefits to a naturally regenerated forest managed for wood production, on a similar rotation. This is especially true if native species are used. In the case of exotic species, the habitat can be improved significantly if the impact is mitigated by measures such as leaving blocks of native species in the plantation, or retaining corridors of natural forest. In Brazil, similar measures are required by government regulation

Sugar

Sugar plantations were highly valued in the Caribbean by the British and French colonists in the 17th and 18th centuries and the use of sugar in Europe rose during this period. Sugarcane is still an important crop in Cuba. Sugar plantations also arose in countries such as Barbados and Cuba because of the natural endowments that they had. These natural endowments included soil that was conducive to growing sugar and a high marginal product of labour realized through the increasing number of slaves.

Rubber

Sugarcane plantation in rural Cuba

Plantings of para rubber, the tree *Hevea brasiliensis*, are usually called plantations.

Oil Palm

Oil palm agriculture is rapidly expanding across wet tropical regions, and is usually developed at plantation scale.

Orchards

Fruit orchards are sometimes considered to be plantations.

Arable Crops

These include tobacco, sugarcane, pineapple, and cotton, especially in historical usage.

Before the rise of cotton in the American South, indigo and rice were also sometimes called plantation crops.

Harvesting tea in Bogor, West Java

Fishing

When Newfoundland was colonized by England in 1610, the original colonists were called “Planters” and their fishing rooms were known as “fishing plantations”. These terms were used well into the 20th century.

The following three plantations are maintained by the Government of Newfoundland and Labrador as provincial heritage sites:

- Sea-Forest Plantation was a 17th-century fishing plantation established at Cuper’s Cove (present-day Cupids) under a royal charter issued by King James I.
- Mockbeggar Plantation is an 18th-century fishing plantation at Bonavista.
- Pool Plantation a 17th-century fishing plantation maintained by Sir David Kirke and his heirs at Ferryland. The plantation was destroyed by French invaders in 1696.

Other fishing plantations:

- Bristol’s Hope Plantation, a 17th-century fishing plantation established at Harbour Grace, created by the Bristol Society of Merchant-Adventurers.
- Benger Plantation, an 18th-century fishing plantation maintained by James Benger and his heirs at Ferryland. It was built on the site of Georgia plantation.
- Piggeon’s Plantation, an 18th-century fishing plantation maintained by Ellias Piggeon at Ferryland.

Slavery and Para-slavery

1913 photo: blacks picking cotton on a plantation in the South

Italian immigrants working on Brazilian coffee plantation, early 20th century

Slave labor extracted from imported Africans was used extensively to work on early plantations (such as tobacco, rice, cotton, and sugar plantations) in the American colonies and the United States, throughout the Caribbean, the Americas, and in European-occupied areas of Africa. Several notable historians and economists such as Eric Williams, Walter Rodney, and Karl Marx contend that the global capitalist economy was largely founded upon the creation and produce of thousands of slave labour camps based in colonial plantations, exploiting tens of millions of purchased Africans.

In modern times, the low wages typically paid to plantation workers are the basis of plantation profitability in some areas. Sugarcane plantations in the Caribbean and Brazil, worked by slave labour, were also examples of the plantation system.

In more recent times, overt slavery has been replaced by *para-slavery* or *slavery-in-kind*, including the sharecropping system. At its most extreme, workers are in "debt bondage": they must work to pay off a debt at such punitive interest rates that it may never be paid off. Others work unreasonably long hours and are paid subsistence wages that (in practice) may only be spent in the company store.

In Brazil, a sugarcane plantation was termed an *engenho* ("engine"), and the 17th-century English usage for organized colonial production was "factory." Such colonial social and economic structures are discussed at Plantation economy.

Sugar workers on plantations in Cuba and elsewhere in the Caribbean lived in company towns known as *bateyes*.

Antebellum American South

In the American South, antebellum plantations were centered on a "plantation house," the residence of the owner, where important business was conducted. Slavery and plantations had different characteristics in different regions of the South. As the Upper South of the Chesapeake Bay colonies developed first, historians of the antebellum South defined planters as those who held 20 or more slaves. Major planters held many more, especially in the Deep South as it developed. The majority of slaveholders held 10 or fewer slaves, often just a few to labor domestically. By the late 18th century, most planters in the Upper South had switched from exclusive tobacco cultivation to mixed-crop production, both because tobacco had exhausted the soil and because of changing markets. The shift away from tobacco meant they had slaves in excess of the number needed for labor, and they began to sell them in the internal slave trade.

There was a variety of domestic architecture on plantations. The largest and wealthiest planter families, for instance, those with estates fronting on the James River in Virginia, constructed mansions in brick and Georgian style, e.g. Shirley Plantation. Common or smaller planters in the late 18th and 19th century had more modest wood frame buildings, such as Southall Plantation in Charles City County.

Old Jackson Plantation house, Schriever, Louisiana, June 1940

In the Lowcountry of South Carolina, by contrast, even before the American Revolution, planters holding large rice plantations typically owned hundreds of slaves. In Charleston and Savannah, the elite also held numerous slaves to work as household servants. The 19th-century development of the Deep South for cotton cultivation depended on large plantations with much more acreage than was typical of the Upper South; and for labor, planters held hundreds of slaves.

Until December 1865 slavery was legal in parts of the United States. Most slaves were employed in agriculture, and *planter* was a term commonly used to describe a farmer with many slaves.

The term planter has no universally-accepted definition, but academic historians have defined it to identify the elite class, "a landowning farmer of substantial means." In the "Black Belt" counties of Alabama and Mississippi, the terms "planter" and "farmer" were often synonymous. Robert Fogel and Stanley Engerman define large planters as owning over 50 slaves, and medium planters as owning between 16 and 50 slaves.

In his study of Black Belt counties in Alabama, Jonathan Wiener defines planters by ownership of real property, rather than of slaves. A planter, for Wiener, owned at least $10,000 worth of real estate in 1850 and $32,000 worth in 1860, equivalent to about the top 8 percent of landowners. In his study of southwest Georgia, Lee Formwalt also defines planters in size of land holdings rather than slaves. Formwalt's planters are in the top 4.5 percent of land owners, translating into real estate worth $6,000 or more in 1850, $24,000 or more in 1860, and $11,000 or more in 1870. In his study of Harrison County, Texas, Randolph B. Campbell classifies large planters as owners of 20 slaves, and small planters as owners of between ten and 19 slaves. In Chicot and Phillips counties, Arkansas, Carl H. Moneyhon defines large planters as owners of twenty or more slaves, and six hundred or more acres.

Community Forestry

Community forestry is an evolving branch of forestry whereby the local community plays a significant role in forest management and land use decision making by themselves in the facilitating support of government as well as change agents. It involves the participation and collaboration of various stakeholders including community, government and non-government organisations (NGO's). The level of involvement of each of these groups is dependent on the specific community forest project, the management system in use and the region. It gained prominence in the mid-1970s and examples of community forestry can now be seen in many countries including Nepal, Indonesia, Korea, Brazil, India and North America.

National Forest Inventory of Peru

Overview

Community forestry is a branch of forestry that deals with the communal management of forests for generating income from timber and non-timber forest products as forms of goods while in other hand regulating ecosystem, downstream settlements benefits from watershed conservation, carbon sequestration and aesthetic values as in forms of services . It has been considered one of the most promising options of combining forest conservation with rural development and community empowerment and poverty reduction objectives. Community forestry is defined by the Food and Agricultural Organization of the United Nations as "any situation that intimately involves local people in forestry activity". Community forestry exists when the local community in an area plays a significant role in land use decision-making and when the community is satisfied with its involvement and benefits from the management of the surrounding forest and its resources.

Community forestry is first implemented through the establishment of a legal and institutional framework including the revision of legal norms and regulations for forest management, the development of National Forest Plans and the strengthening of decentralization processes to sub-national levels of government. The second principal line of action is the implementation of pilot projects to demonstrate the feasibility of the community forestry framework. However, a study by the Overseas Development Institute shows that the technical, managerial and financial requirements stipulated by the framework are often incompatible with local realities and interests. A successful legal and institutional framework will incorporate the strengthening of existing institutions and enable the dissemination of locally appropriate practices as well as the local capacity for regulation and control.

In a 2016 review of community-based forestry, FAO estimated that almost one-third of the world's forest area is under some form of community-based management.

History

Community forestry first came to prominence in the mid-1970s and has continued to evolve over the last few decades in a growing number of countries. The availability of forest resources are often greatly reduced for use by the local people due to increasing pressures to cultivate the land, reliance on the forest resources are also affected by economic and political changes. The evolution of community forestry in Nepal dates back to the late 1970s and was first instilled as an attempt to improve the management of forest resources and address environmental issues that were of great concern with the countries failing centralized forest policy. Over the past two decades, community forestry has been applied successfully in many developing countries, with its main goal being the alleviation of poverty amongst local forest communities and forest conservation. More recently, community forestry has been implemented in developing countries and it has been successful in its aims of sustainable forest management, climate change adaptation plan of action, and securing socio-economic benefits for local communities.

Stakeholders

There is a large variety of stakeholders involved when considering community forestry. Participation from some of the various levels of community, government and non-government organisations (NGO's) are essential in the project's success. While specific stakeholders vary between different community forestry projects the primary stakeholder groups are as follows:

Local Community
Communities living adjacent to or within the forest
Traditional authorities
Community-based organisations (forest users groups etc.)
Community representatives / local councils
Government
State Government
National Government
Departments of Agriculture, Forestry, Environment etc. (specific the country and/or region)
NGO's
Environmental/conservation groups
Commercial forestry industries

Industries reliant on forests (i.e. harvestable products other than wood, e.g. game meat)
Tourism industry
Animal welfare groups
Natural Ecosystem

Table 1. Common stakeholders involved with community forestry.

Stakeholders of community forestry have a vested interest to establish sustainable practices, whether this be to develop and maintain a regular income, ensure that forests are sufficiently protected to ensure their longevity or to reduce illegal activities and manage the area in such a way to promote tourism and conservation. In this situation, stakeholders came to conclusion to handover forest resources to local communities for conserving, managing and utilization by their own decision. Despite significant development, continued improvement in the collaboration between local governments and forest communities seems to be a key point for better community forest management. A wide range of futures scenarios have been put up to help the environmental decision process.

Challenges

A study conducted in the Brazilian Amazon determined that there are a number of challenges that must be faced when developing a sustainable management strategy for community forestry. These challenges are outlined in Table 2 and Figure shows the impact each management challenge has on other obstacles. The model is segregated into two phases: the *development phase* during which several enabling factors (land ownership, organisational capacity, technical knowledge and capital) are needed to obtain a legal management permit and secondly the *operational phase* where factors (clandestine loggers, access to markets, infrastructure and managerial skills) influence the successfulness of the management program. Each of the challenges outlined in Ta-ble 1 must be addressed in order for a self-sustaining community forestry management program to be established.

Land ownership	It is not uncommon for small settlements living traditionally to lack clear title to the forests and lands, as such their rights to harvest the land may come into dispute. Formal recognition of land ownership or rights to use is needed for legal forest management. In Nepal, there is still confusion to communities about their tenure into services such as income from trading of carbon sequestration come from forestry resources. Government wants to hold revenue raised from trading of carbon but communities claim that it is our property due tobe managed by us.
Organizational capacity	Organizational capacity refers to the community's competence to organize a forest management program.
Capital	A community will require start-up capital to invest in the required infrastructure, equipment, and to hire a forester to organise and oversee management plans.

Technical knowledge	While it is common for community members have a thorough understanding of forest ecology in a natural and historical sense, they often lack the technical knowledge and legal certification required to manage forest resources legally.
Legal management	Community forestry needs to be based on a legal management plan, prepared and approved by the relevant government authority (usually state environment agencies). Approval of the management plan can often be a long, bureaucratic process.
Clandestine loggers	Clandestine (illegal) loggers can enter a region and illegally log valuable species without the community's consent, and without a legal management plan. Illegal harvesting often significantly degrades the forest as few limitations are observed by the illegal loggers. Communities are either robbed of their valuable timber, or are paid below market prices.
Market access	Small villages that are often the focus of community forestry initiatives generally have limited access to markets due to their physical isolation, precarious transport and communication, limited contact with buyers and lack of marketing knowledge. Often they will also face difficulties competing with large-scale operations and illegal timber that flood the market with cheaper products.
Infrastructure	Forest management, for all types of enterprise, requires a certain amount of physical infrastructure, such as roads, logging equipment, buildings for storage and management headquarters and/or a reliable power supply.
Managerial skills	Skills related to effective management of the program, business knowhow and entrepreneurial ability, day-to-day decision-making, marketing skills, ability to resolve internal conflicts, and ensuring community benefit sharing are often lacked in rural villages.
Economic returns	The result of inadequacies in management is small and often insufficient economic returns to keep the program afloat, and furthermore to keep the community interested in the activity.

Table 2. Challenges faced by stakeholders managing forests

Best Practices

For a conservation program within a Community forest to succeed, community involvement is imperative. Governments with interest in forest conservation introduce statewide policies and legislations which have historically failed to deliver the desired outcomes such as in China, Nepal and Peru. Moreover, no single stakeholder by itself can ensure the success of such a program.

Community Forestry and Wildlife Conservation

Though there is little research on the role of community forestry to wildlife conservation, some empirical studies suggests that it help in wildife conservation. It is done by decreasing the human disturbance, increasing regeneration of forest and increasing of ground cover.

Nepal

Common land in Nepal is owned by the state which often does not have the capacity to monitor and manage the area. This often leads to the over use of the resources by the community due to lack of incentives. To overcome this, programs involving community participation were introduced and 'Forest User Groups' (FUG) formed to manage the forests resources without giving them ownership of the land. Community forest management system in Nepal becomes one of the successful program out of 8 around world that is recognized on Rio 20+. This has resulted in better outcomes in the region.

Indonesia

Villagers from three communities in Bantaeng district in Indonesia, with the assistance of a funded project obtained a Forest management license and secured a 35-year lease on their local forest. With the incentive to preserve their source of income, the villagers have had a positive impact on forest conservation. This is now viewed as a reference model for the Ministry of Forestry's future projects.

Korea

Farmers enter into a profit sharing contract with the 'Village Forestry Association' (VFA) to form a cooperative which assists farmers with reforestation in keeping with the legislation. VFA, though loosely linked to the Office of Forestry enjoys a degree of autonomy facilitating community participation. This system demonstrates the desirable mix of top-down and bottom-up planning ensuring government control as well as effective reforestation through active community participation.

Management System

As it is unlikely for any two areas to have identical socio-economic and environmental conditions, there is no blueprint management plan that would provide a blanket solution to all the problems. Based on research over several years in Nepal, it was noted that to have and effective management system, we need to identify variables which would affect the success of the system and group them into the following five sets:

(I) attributes of the resource system,

(II) attributes of the user group,

(III) attributes of the governance system,

(IV) attributes relating to interactions between the user group and resource, and

(V) attributes relating to interactions between the governance system and the resource.

In some cases, it is unrealistic to expect progress in a community level management of forest activities, as often conflict arise with respect to land use and benefit sharing within the community. Such issues can be overcome by recognising that a community level of management may not be the most effective management technique. Instead, adopting the following approach will provide a solution -

(I) Management by smaller work groups within the community, sharing common interest in the resources.

(II) A clear management plan with specific benefit sharing arrangement within the work group.

(III) Develop management systems which are within the expertise of the working group.

Faith communities are increasingly participating in efforts to promote ecological sustainability. Whereas the last 50–100 years has seen them lease out their territory to industry, they are beginning to reclaim and restore this land. Their recognition amongst local and national authorities have allowed community forestry schemes to develop and there have been several highly successful efforts across Cambodia, which have been extended to Vietnam and Laos. In September 2010, Buddhists monks were awarded the UNDP's Equator Prize for their ongoing conservation work. This comprises the establishment of tree nurseries, seedling distribution, ordination activities, composting schemes and a vegetable garden. They are using this work to promote sustainable living and conservation throughout three districts; the pilot project, launched in 1997 in Kratie consists of a six pagoda network that covers 1,461 hectares (3,610 acres). The site has become a valuable source of information and environmental education as well as a base for the local community. Villagers regularly hold panel discussions about how they can go on to best take advantage of what the natural forest has to offer without destroying it.

Incentives

Farm Forest for Household Use

(I) Preserving the source of community's livelihood

(II) Maintaining soil integrity by preventing erosion.

(III) Use of fuel wood and fodder- Planting trees to provide fuel wood as well as fodder for cattle can be encouraged in ways that do not conflict with cash crops and food production. e.g. The Neem tree was introduced in West Africa and is now the most widely grown tree in the drier parts of the continent. It was easily cultivated and provides the farmers with good timber, fuel and shade.

(IV) Developing nursery networks through support activities which assist with subsidies. This encourages entrepreneurship to produce tree planting stock for sale.

Farm Forests for Markets

Tree growing can take on the characteristics of a crop where there is a market for wood products such as poles, fuel wood and, pulp for production of paper. Companies tie up with farmers for supply of these products giving a steady source of income to the farming community. For example, in Philippines, over 3000 farmers cultivate trees for pulp production for an industry that provides a market as well as a minimum price for the product, and Cooperatives of Village Forestry Association in Korea have helped local communities cater to a growing market in forest products such as timber and mushrooms.

Forest Farming

Forest farming is the cultivation of high-value specialty crops under a forest canopy that is intentionally modified or maintained to provide shade levels and habitat that favor growth and enhance production levels. Forest farming encompasses a range of cultivated systems from introducing plants into the understory of a timber stand to modifying forest stands to enhance the marketability and sustainable production of existing plants.

Forest farming is a type of agroforestry practice characterized by the "four I's": intentional, integrated, intensive and interactive. Agroforestry is a land management system that combines trees with crops or livestock, or both, on the same piece of land. It focuses on increasing benefits to the landowner as well as maintaining forest integrity and environmental health. The practice involves cultivating non-timber forest products or niche crops, some of which, such as ginseng or shiitake mushrooms, can have high market value.

Non-timber forest products (NTFPs) are plants, parts of plants, fungi, and other biological materials harvested from within and on the edges of natural, manipulated, or disturbed forests. Examples of crops are ginseng, shiitake mushrooms, decorative ferns, and pine straw. Products typically fit into the following categories: edible, medicinal and dietary supplements, floral or decorative, or specialty wood-based products.

History

Forest farming, though not always by that name, is practiced around the world. For centuries, humans have relied on fruits, nuts, seeds, parts of foliage and pods from trees and shrubs in the forests to feed themselves and their livestock. Over time, certain species have been selected for cultivation near homes or livestock to provide food or medicine. For example, in the southern United States, mulberry trees are used as a feedstock for pigs and often cultivated near pig quarters.

Toyohiko Kagawa, forest farming pioneer.

In 1929, J. Russell Smith, Emeritus Professor of Economic Geography at Columbia University, published "Tree Crops – A Permanent Agriculture" which stated that crop-yielding trees could provide useful substitutes for cereals in animal feeding programs, as well as conserve environmental health. Toyohiko Kagawa read and was heavily influenced by Smith's publication and began experimental cultivation under trees in Japan during the 1930s. Through forest farming, or three-dimensional forestry, Kagawa addressed problems of soil erosion by persuading many of Japan's upland farmers to plant fodder trees to conserve soil, supply food and feed animals. He combined extensive plantings of walnut trees, harvested the nuts and fed them to the pigs, then sold the pigs as a source of income. When the walnut trees matured, they were sold for timber and more trees were planted so that there was a continuous cycle of economic cropping that provided both short-term and long-term income to the small landowner. The success of these trials prompted similar research in other countries. Unfortunately, World War II disrupted communication and slowed advances in forest farming. In the mid-1950s research resumed in places such as southern Africa. Kagawa was also an inspiration to Robert Hart pioneered forest gardening in temperate climates in the sixties in Shropshire, England.

In earlier years, livestock were often considered part of the forest farming system. Now they are typically excluded and agroforestry systems that integrate trees, forages and livestock are referred to as silvopastures. Because forest farming combines ecological stability of natural forests and productive agriculture systems, it is considered to have great potential for regenerating soils, restoring ground water supplies, controlling floods and droughts and cultivating marginal lands. In addition to these benefits for re-establishing productive forests on marginal lands, forest farming is way to add financial value while conserving land that is currently forested, as discussed in the methods section.

In more recent years, there has been growing interest in locally grown and organic foods throughout the United States. There has been an increase in farmer's markets and community-supported agriculture small enterprises. These have also become outlets for NTFPs. In order to stay competitive, many farmers look to add unique crops to their product line. With the quantity and quality of resources developing online that offer tutorials and educational information on how to create and maintain forest farms, forest gardens, how to cultivate specific crops such as shiitake mushrooms and how to successfully market these items, forest farming is expanding as a viable land management practice. Good places to look for research-based resources are the USDA National Agroforestry Center's publication section, the Center for Agroforestry at the University of Missouri, the Cornell Cooperative Extension, the Non-Timber Forest Products website by The Virginia Tech Department of Wood Science and Forest Products, the USDA Forest Service Southern Research Station and the Top of the Ozarks RC&D in Missouri and the collaborative Forest Farming community of practice on eXtension.org, the online presence of the Cooperative Extension System of the US Land Grant Universities.

Principles

Forest farming principles constitute an ecological approach to forest management. Forest resources are judiciously used while biodiversity and wildlife habitat are conserved. Forest farms have the potential to restore ecological balance to fragmented second growth forests through intentional manipulation to create the desired forest ecosystem.

In some instances, the intentional introduction of species for botanicals, medicinals, food or decorative products is accomplished using existing forests. The tree cover, soil type, water supply, land form and other site characteristics determine what species will thrive. Developing an understanding of species/site relationships as well as understanding the site limitations is necessary to utilize these resources for production needs, while conserving adequate resources for the long-term health of the forest.

Apart from the environmental benefits, forest farming can increase the economic value of forest property and provide short- and long-term benefits to the landowner. Forest farming provides economic return from intact forest ecosystems, but timber sales can remain part of the long-term management strategy.

Methods

Forest farming methods may include: Intensive, yet careful thinning of overstocked, suppressed tree stands; multiple integrated entries to accomplish thinning so that systemic shock is minimized; and interactive management to maintain a cross-section of healthy trees and shrubs of all ages and species. Physical disturbance to the surrounding area should be minimized. The following are forest farming techniques described in the Training Manual produced by the Center for Agroforestry at the University of Missouri.

Level of Management Required

1. Forest gardening is the most intensive of forest farming methods. In addition to thinning the overstory, this method involves clearing the understory of undesirable vegetation and other practices that are closely related to agronomy (tillage, fertilization, weeding, and control of disease and insects and wildlife management). Due to input levels, this method often produces lower valued products compared to other methods. Forest gardens take advantage of the vertical levels of light availability and space under the forest canopy so that more than one crop can be grown at once if desired.

2. Wild-simulated seeks to maintain a natural growing environment, yet enriches local NTFP populations to create an abundant renewable supply of the products. Minimal disturbance and natural growing conditions ensure products will be similar in appearance and quality of those harvested from the wild. Rather than till, practitioners often rake leaves to expose soil, sow seed directly onto the ground, and then cover with leaves again. Since this method produces NTFPs that closely resemble wild plants; they often command a higher price than NTFPs produced using the forest gardening method.

3. Forest tending involves adjusting tree crown density to manipulate light levels that favor natural reproduction of desirable NTFPs. This low intensity management approach does not involve supplemental planting to increase populations of desired NTFPs.

4. Wildcrafting is the harvesting of naturally growing NTFPs. It is not considered a forest farming practice since there is no human involvement in the plant's establishment and maintenance. However, wildcrafters often take steps to protect NTFPs with future harvests in mind. It becomes agroforestry once forest thinnings, or other inputs, are applied to sustain or maintain plant populations that might otherwise succumb to successional changes in the forest. The most important difference between forest farming and wildcrafting is that forest farming intentionally produces NTFPS, whereas wildcrafting seeks and gathers from naturally growing NTFPs.

Production Considerations

Forest farming can be a small business opportunity for landowners and requires careful planning, including a business and marketing plan. Learning how to market the NTFPs on the Internet is an option, but may entail higher shipping costs. Landowners should consider all options for selling their products including, farmer's markets or restaurants that focus on locally grown ingredients. The development phase should include a forest management plan that states the landowner's objectives and a resource inventory. Start-up costs should be analyzed as specific equipment may be necessary to harvest or process the product, whereas other crops require minimal initial investment. Local incentives for sustainable forest management, as well as regulations and policies should be explored. The Convention on International Trade in Endangered Species of Wild Fauna and Flora (CITES) regulates international trade of certain plant (American ginseng and goldenseal) and animal species. To be legally exported, regulated plants must be harvested and records kept according to CITES rules and restrictions. Many states also have harvesting regulations for certain native plants that are searchable online. Another good source to start with on information is the Medicinal Plants at Risk 2008 report, by the Center for Biological Diversity] in the U.S.

Examples of Crops

(from the National Agroforestry Center)

Medicinal herbs:

- Ginseng (*Panax quinquefolius*)
- Black Cohosh (*Actaea racemosa*)
- Goldenseal (*Hydrastis canadensis*)
- Bloodroot (*Sanguinaria canadensis*)
- Pacific yew (*Taxus brevifolia*)
- Mayapple (*Podophyllum peltatum*)
- Saw palmetto (*Serenoa repens*)
- American Pokeweed (*Phytolacca americana*)

Nuts:

- Black walnut (*Juglans nigra*)
- Hazelnut (*Corylus avellana*)
- Shagbark hickory (*Carya ovata*)
- Beechnut (*Fagus sylvatica*)

Fruit:

- Pawpaw (*Asimina triloba*)
- Currants (*Ribes spp*)
- Elderberry (*Sambucus spp*)
- Serviceberry (*Amelanchier spp*)
- Blackberry (*Rubus spp*)
- Huckleberry (*Gaylussacia brachycera*)

Other food crops:

- Ramps (wild leeks) (*Allium tricoccum*)
- Syrups (maple)
- Honey
- Mushrooms
- Other edible roots

Other products: (mulch, decoratives, crafts, dyes)

- Pine straw
- Willow twigs
- Vines
- Beargrass (*Xerophyllum tenax*)
- Ferns
- Pine cones
- Moss

Native ornamentals:

- Rhododendron (*Rhododendron catawbiense*)
- Highbush cranberry (*Viburnum trilobum*)
- Flowering dogwood (*Cornus florida*)

High Conservation Value Forest

High conservation value forest (HCVF) is a Forest Stewardship Council (FSC) forest management designation used to describe those forests who meet criteria defined by the FSC Principles and Criteria of Forest Stewardship.

Specifically, high conservation value forests are those that possess one or more of the following attributes:

- forest areas containing globally, regionally or nationally significant: concentrations of biodiversity values (e.g. endemism, endangered species, refugia); and/or large landscape-level forests, contained within, or containing the management unit, where viable populations of most if not all naturally occurring species exist in natural patterns of distribution and abundance

- forest areas that are in or contain rare, threatened or endangered ecosystems
- forest areas that provide basic services of nature in critical situations (e.g. watershed protection, erosion control)
- forest areas fundamental to meeting basic needs of local communities (e.g. subsistence, health) and/or critical to local communities' traditional cultural identity (areas of cultural, ecological, economic or religious significance identified in cooperation with such local communities).

History

The arose relatively recently in connection with the development of standards for the certification of forest management. The term was developed by the Forest Stewardship Council, A.C. (FSC), which is an international accreditation association1 incorporated in 1995. The FSC first started considering the term ***high conservation values*** in 1996 and formally included the term *high conservation value forests* in the FSC Principles and Criteria of Forest Stewardship in January 1999.

A number of terms have been used somewhat interchangeably with the term HCVF. To a certain extent this arises from translation between languages and to a certain extent it arises from varying approaches at interpretation. The Spanish version uses the terms *bosques con alto valor de conservación* (forests with high value of conservation) and *bosques con alto valor para la conservación* (forests with high value for conservation).

As noted, the term HCVF arose following consideration of the term *high conservation values*; this usage led to the concept of *forests having high conservation values* and *forests with high conservation values*. Indeed, the term *high conservation value* is sometimes given its own acronym HCV; however, the FSC refers to an HCV as "attributes".

Examples

Examples of the large landscape globally important forests include the Canada boreal forest and the remaining Amazon rainforest.

Examples of forest types which embrace rare or endangered species include the Dry deciduous forests of Madagascar the two Monterey Cypress of central coastal California and the Maritime Coast Range Ponderosa Pine forests of coastal California.

HCVF Toolkits

Following the inclusion of the term in the FSC P&C in 1999, the FSC formed an Advisory Panel for the Implementation of high conservation value forests and the precautionary principle. This panel produced a number of consultation documents on how to identify HCVF, including indicators and verifiers. However, this consultation process was not finalized and FSC has yet to issue any specific guidance regarding HCVF1.

In lieu of formal guidance from FSC, there has been a proliferation of discussion on HCVF, led mainly by non-governmental organisations (NGOs). In particular, the need for more precise and practical guidance led to a 2001 initiative by the UK-based consulting company, Proforest Ltd. This initiative led to the production of a three-part High Conservation Value Toolkit (the 'Global Toolkit'). The Global Toolkit offered a revised definition of HCVF as follows:

A High Conservation Value Forest is the area of forest required to maintain or enhance a High Conservation Value.

This definition provides a new approach to the zonation of HCVF as well as introduces the concept of 'High Conservation Value'. The inclusion of the term 'area of forest' provide clarity that there may be instances where an HCVF zone might be restricted to part of a forest, while the FSC definition implied that the presence of one or more HCV attribute would render the whole forest as to be a 'high conservation value forest'. Indeed, despite the provision for partial zonation, the Global Toolkit makes it clear that for the purposes of certification, any forest management unit containing even one small HCV is classified as a High Conservation Value Forest, which brings into play extra requirements for consultation, maintenance or enhancement, and annual monitoring. However, the additional burden is determined by the nature of the HCV and will affect only those parts of the forest, and those aspects of management, that are required for the maintenance or enhancement of the HCV (this procedure is explained in more detail in the section on Adaptive Management, below).

There are six recognized forms of High conservation values forests:

HCV1. Forest areas containing globally, regionally or nationally significant concentrations of biodiversity values (e.g. endemism, endangered species, refugia).

HCV2. Forest areas containing globally, regionally or nationally significant large landscape level forests, contained within, or containing the management unit, where viable populations of most if not all naturally occurring species exist in natural patterns of distribution and abundance.

HCV3. Forest areas that are in or contain rare, threatened or endangered ecosystems.

HCV4. Forest areas that provide basic services of nature in critical situations (e.g. watershed protection, erosion control).

HCV5. Forest areas fundamental to meeting basic needs of local communities (e.g. subsistence, health).

HCV6. Forest areas critical to local communities' traditional cultural identity (areas of cultural, ecological, economic or religious significance identified in co-operation with such local communities).

The Global Toolkit has not been explicitly endorsed by the FSC but the revised ordering system appears to have gained widespread appeal and there is a recommendation for its use in FSC reports found in a footnote to an FSC reporting standard. The Global Toolkit was intentionally designed for a wide range of users beyond the FSC scheme. Indeed, in December 2005, the popularity of the HCV approach led to the formation of the HCV Resource Network which includes a broad platform of NGOs as well as the World Bank and the International Tropical Timber Organization (ITTO).

High Conservation Value Areas

The concept of a High conservation value forest has now been generalized and used by other certification schemes like the Roundtable on Sustainable Palm Oil (RSPO), the Better Cotton Initiative, the Sustainable Agriculture Network (SAN) and Bonsucro, among others. Consumer goods companies (mainly members of the Consumer Goods Forum) have also made independent commitments to protecting High Conservation Values in their supply chains.

The analagous six High Conservation Values are:

- HCV 1: Concentrations of biological diversity including endemic species, and rare, threatened or endangered species, that are significant at global, regional or national levels.
- HCV 2: Intact forest landscapes and large landscape-level ecosystems and ecosystem mosaics that are significant at global, regional or national levels, and that contain viable populations of the great majority of the naturally occurring species in natural patterns of distribution and abundance.
- HCV 3: Rare, threatened, or endangered ecosystems, habitats or refugia.
- HCV 4: Basic ecosystem services in critical situations, including protection of water catchments and control of erosion of vulnerable soils and slopes.
- HCV 5: Sites and resources fundamental for satisfying the basic necessities of local communities or indigenous peoples (for livelihoods, health, nutrition, water, etc...), identified through engagement with these communities or indigenous peoples.
- HCV 6: Sites, resources, habitats and landscapes of global or national cultural, archaeological or historical significance, and/or of critical cultural, ecological, economic or religious/sacred importance for the traditional cultures of local communities or indigenous peoples, identified through engagement with these local communities or indigenous peoples..

References

- Matta, Rao (2015). Towards effective national forest funds, FAO Forestry Paper 174 (PDF). Rome, Italy: Food and Agriculture Organization of the United Nations. ISBN 978-92-5-108706-0.
- Kohm, K. A, and Franklin, J. F., Creating a forestry for the 21st century: the science of ecosystem management. Island Press. 1997, ISBN 978-1-55963-399-4
- Don, Gilmour (2016). FAO Forestry Paper 176: Forty years of community-based forestry: A review of its extent and effectiveness (PDF). FAO. ISBN 978-92-5-109095-4. Retrieved 5 April 2016.
- Barnard, Gerald Foley, Geoffrey (1984). Farm and community forestry. London: Earthscan, International Institute for Environment and Development. ISBN 0-905347-53-6.
- Helms, John A. (1998-09-01). The Dictionary of Forestry. Society of American Foresters. ISBN 978-0-939970-73-5.
- Savill, P. Evans, J. Auclair, D. Falk, J. (1997). Plantation Silviculture in Europe. Oxford University Press. Oxford. ISBN 0-19-854909-1
- "LEDS GP Agriculture, Forestry and Other Land Use Working Group factsheet" (PDF). Low Emission Development Strategies Global Partnership (LEDS GP). Retrieved 23 March 2016.
- "Sustainable Forest Management Toolbox" (PDF). Food and Agriculture Organization of the United Nations. Retrieved 24 June 2014.

4

Understanding Silviculture

Forests need to be surveyed and monitored and their growth and health needs. Some of the aspects of silviculture explained in the following chapter are tree planting, plant nursery, thinning, pruning and clearcutting. The topics discussed in the text are of great importance to broaden the existing knowledge on silviculture.

Silviculture

Silviculture is the practice of controlling the establishment, growth, composition, health, and quality of forests to meet diverse needs and values.

The name comes from the Latin *silvi-* (forest) + culture (as in growing). The study of forests and woods is termed silvology. Silviculture also focuses on making sure that the treatment(s) of forest stands are used to preserve and to better their productivity.

Generally, silviculture is the science and art of growing and tending forest crops, based on a knowledge of silvics, i.e., the study of the life history and general characteristics of forest trees and stands, with particular reference to locality factors. More particularly, silviculture is the theory and practice of controlling the establishment, composition, constitution, and growth of forests. No matter how forestry as a science is constituted, the kernel of the business of forestry is silviculture, as it includes direct action in the forest, and in it all economic objectives and technical considerations ultimately converge. The kernel of silviculture is regeneration.

Suggestions for how best to go about the job, presented by Jeglum et al. (2003), though aimed primarily at the boreal forest in Ontario, merit wider consideration. The 110-page publication describes Best Management Practices, first by general principles, then by sensitive sites. Illustrations are plentiful and are well chosen to complement this excellent text.

To some the distinction between forestry and silviculture is that silviculture is applied at the stand level and forestry is broader. For example, John D. Matthews says "complete regimes for regenerating, tending, and harvesting forests" are called "silvicultural systems".

So adaptive management is required for silviculture, whereas forestry can be natural, conserved land without a stand level treatment being applied. A common taxonomy divides silviculture into regenerating, tending and harvesting techniques.

An employee of the White Mountain Apache Tribe forestry department uses a hoedad to dig a hole for a ponderosa pine seedling in the shade of a burned tree stump, which protects the seedling from windy conditions that dry the trees

Silvicultural Systems

The origin of forestry in German speaking Europe has defined silvicultural systems broadly as Hochwald - High forest, Mittelwald - Coppice with standards (and Compound coppice), Short rotation coppice and Niederwald - Coppice. There are other systems too. These varied silvicultural systems include several harvesting methods, which are often wrongly said to be a silvicultural systems, but may also be called as rejuvenating or regenerating method depending on the purpose.

Hochwald is further subdivided in German into: Altersklassenwald - Age class forest which includes:

- *even aged forestry*
    - - Kahlschlag - Clear cutting
    - - Schirmschlag - Shelterwood cutting
    - - Seed-tree method
- *un-even aged forestry*
    - - Femelschlag (Femelschlagverfahren) - The Femel selection cutting (Group selection cutting),
    - - Saumschlag (Saumfemelschlagbetrieb) - Strip selection cutting (strip-and-group felling system)
    - - Schirmkeilschlag - Shelterwood wedge cutting
    - - Mischformen - mixed form regeneration methods.

and Dauerwald - Continuous cover forestry, in which are found:

- *un-even aged forestry*

- - Plenterwald - Selection forest
- - Zielstärkennutzung - Target diameter harvesting.

From these names the impression is given that these are neatly defined systems, but in practice there are a variety of variations within these harvesting methods according to local ecology and site conditions. While an individual location can be identified where it could be said to be the archetypal form of harvesting technique (they all originated somewhere with a particular forester, and have been described in the scientific literature), and broad generalizations can be made these are merely rules of thumb rather than strict blue prints on how techniques might be applied. This misunderstanding has meant that many older English textbooks did not capture the true complexity of silviculture as practiced where it originated in Mitteleuropa.

This silviculture was culturally predicated on wood production in temperate and boreal climates and did not deal with tropical forestry. The misapplication of this philosophy to those tropical forests has been problematic. There is also an alternative silvicultural tradition which developed in Japan and thus created a different biocultural landscape called satoyama.

After harvesting comes regeneration, which may be split into Natural and Artificial. And tending which includes release treatments, pruning, thinning and intermediate treatments. As it can be imagined it maybe that any of these 3 phases (harvesting - regeneration - tending) may happen at the same time, within a stand, depending on the goal for that particular stand.

Regeneration

Regeneration is basic to the continuation of forested, as well as to the afforestation of treeless land. Regeneration can take place through self-sown seed ("natural regeneration"), by artificially sown seed, or by planted seedlings. In whichever case, the performance of regeneration depends on its growth potential and the degree to which its environment allows the potential to be expressed. Seed, of course, is needed for all regeneration modes, both for natural or artificial sowing and for raising planting stock in a nursery.

Natural regeneration is:

"Human-assisted natural regeneration" means establishment of a forest age class from natural seeding or sprouting in an area after harvesting in that area through selection cutting, shelter (or seed-tree) harvest, soil preparation, or restricting the size of a clearcut stand to secure natural regeneration from the surrounding trees.

The process of natural regeneration involves the renewal of forests by means of self-sown seeds, root suckers, or coppicing. In natural forests, conifers rely almost entirely on regeneration through seed. Most of the broadleaves, however, are able to regenerate by the means of emergence of shoots from stumps (coppice) and broken stems.

Seedbed Requirements

Any seed, self-sown or artificially applied, requires a seedbed suitable for securing germination.

In order to germinate, a seed requires suitable conditions of temperature, moisture, and aeration. For seeds of many species, light is also necessary, and facilitates the germination of seeds in other species, but spruces are not exacting in their light requirements, and will germinate without light. White spruce seed germinated at 35 °F (1.7 °C) and 40 °F (4.4 °C) after continuous stratification for one year or longer and developed radicles < 6 cm long in the cold room. When exposed to light, those germinants developed chlorophyll and were normally phototropic with continued elongation.

For survival in the short and medium terms, a germinant needs: a continuing supply of moisture; freedom from lethal temperature; enough light to generate sufficient photosynthate to support respiration and growth, but not enough to generate lethal stress in the seedling; freedom from browsers, tramplers, and pathogens; and a stable root system. Shade is very important to the survival of young seedlings. In the longer term, there must be an adequate supply of essential nutrients and an absence of smothering.

In undisturbed forest, decayed windfallen stemwood provides the most favourable seedbed for germination and survival, moisture supply being dependable, and the elevation of seedlings somewhat above the general level of the forest floor reduces the danger of smothering by leaves and snow-pressed minor vegetation; nor is such a microsite likely to be subject to flooding. Advantages conferred by those microsites include: more light, higher temperatures in the rooting zone, and better mycorrhizal development. In a survey in the Porcupine Hills, Manitoba, 90% of all spruce seedlings were rooted in rotten wood.

Mineral soil seedbeds are more receptive than the undisturbed forest floor, and are generally moister and more readily rewetted than the organic forest floor. However, exposed mineral soil, much more so than organic-surfaced soil, is subject to frost heaving and shrinkage during drought. The forces generated in soil by frost or drought are quite enough to break roots.

The range of microsites occurring on the forest floor can be broadened, and their frequency and distribution influenced by site preparation. Each microsite has its own microclimate. Microclimates near the ground are better characterized by vapour pressure deficit and net incident radiation, rather than the standard measurements of air temperature, precipitation, and wind pattern.

Aspect is an important component of microclimate, especially in relation to temperature and moisture regimes. Germination and seedling establishment of Engelmann spruce were much better on north than on south aspect seedbeds in the Fraser Experimental Forest, Colorado; the ratios of seeds to 5-year-old seedlings were determined as 32:1,

76:1, and 72:1 on north aspect bladed-shaded, bladed-unshaded, and undisturbed-shaded seedbeds, respectively. Clearcut openings of 1.2 ha to 2.0 ha adjacent to an adequate seed source, and not more than 6 tree heights wide, could be expected to secure acceptable regeneration (4,900, 5-year-old trees/ha), whereas on undisturbed-unshaded north aspects, and on all seedbed treatments tested on south aspects, seed to seedling ratios were so high that the restocking of any clearcut opening would be questionable.

At least 7 variable factors may influence seed germination: (1) seed characteristics, (2) light, (3) oxygen, (4) soil reaction (ph), (5) temperature, (6) moisture, and (7) seed enemies. Moisture and temperature are the most influential, and both are affected by exposure. The difficulty of securing natural regeneration of Norway spruce and Scots pine in northern Europe led to the adoption of various forms of reproduction cuttings that provided partial shade or protection to seedlings from hot sun and wind. The main objective of echeloned strips or border-cuttings with northeast exposure was to protect regeneration from overheating, and was originated in Germany and deployed successfully by Alarik (1925) and others in Sweden. On south and west exposures, direct insolation and heat reflected from tree trunks often result in temperatures lethal to young seedlings, as well as desiccation of the surface soil, which inhibits germination. The sun is less injurious on eastern exposures because of the lower temperature in the early morning, related to higher humidity and presence of dew.

Baldwin (1933), after noting that summer temperatures in North America are often higher than those in places where border-cuttings have been found useful, reported the results of a survey of regeneration in a stand of red spruce plus scattered white spruce that had been isolated by clearcutting on all sides, so furnishing an opportunity for observing regeneration on different exposures in this old-field stand at Dummer, New Hampshire. The regeneration included a surprisingly large number of balsam fir seedlings from the 5% stand component of that species. The maximum density of spruce regeneration, determined 4 rods (20 m) inside from the edge of the stand on a North 20°E exposure, was 600,000/ha, with almost 100,000 balsam fir seedlings.

A prepared seedbed remains receptive for a relatively short period, seldom as long as 5 years, sometimes as short as 3 years. Seedbed receptivity on moist, fertile sites decreases with particular rapidity, and especially on such sites, seedbed preparation should be scheduled to take advantage of good seed years. In poor seed years, site preparation can be carried out on mesic and drier sites with more chance of success, because of the generally longer receptivity of seedbeds there than those on moister sites. Although an indifferent seed year can suffice if seed distribution is good and environmental conditions favourable to seedling germination and survival, small amounts of seed are particularly vulnerable to depredation by small mammals. Considerable flexibility is possible in timing site preparation to coincide with cone crops. Treatment can be apllied either before any logging takes place, between partial cuts, or after logging. In cut and leave strips, seedbed preparation can be carried out as a single operation, pre-scarifying the leave strips, post-scarifying the cut strips.

Broadcast burning is not recommended as a method of preparing sites for natural regeneration, as it rarely exposes enough mineral soil to be sufficiently receptive, and the charred organic surfaces are a poor seedbed for spruce. A charred surface may get too hot for good germination and may delay germination until fall, with subsequent overwinter mortality of unhardened seedlings. Piling and burning of logging slash, however, can leave suitable exposures of mineral soil.

Season of Planting

Artificial Regeneration

With a view to reducing the time needed to produce planting stock, experiments were carried out with white spruce and three other coniferous species from Wisconsin seed in the longer, frost-free growing season in Florida, 125 vs. 265 days in central Wisconsin and northern Florida, respectively. As the species studied are adapted to long photoperiods, extended daylengths of 20 hours were applied in Florida. Other seedlings were grown under extended daylength in Wisconsin and with natural daylength in both areas. After two growing seasons, white spruce under long days in Florida were about the same as those in Wisconsin, but twice as tall as plants under natural Wisconsin photoperiods. Under natural days in Florida, with the short local photoperiod, white spruce was severely dwarfed and had a low rate of survival. Black spruce responded similarly. After two growing seasons, long day plants of all 4 species in Florida were well balanced, with good development of both roots and shoots, equaling or exceeding the minimum standards for 2+1 and 2+2 outplanting stock of Lake States species. Their survival when lifted in February and outplanted in Wisconsin equalled that of 2+2 Wisconsin-grown transplants. Artificial extension of the photoperiod in the northern Lake States greatly increased height increment of white and black spruces in the second growing season.

Optimum conditions for seedling growth have been determined for the production of containerized planting stock. Alternating day/night temperatures have been found more suitable than a constant temperature; at 400 lumens/m^2 light regime, a 28 °C/20 °C day/night temperatures have been recommended for white spruce. However, temperature optima are not necessarily the same at different ages and sizes. Tinus (1984) investigated the effects of combinations of day and night temperature on height, caliper, and dry weight of 4 seed sources of Engelmann spruce. The 4 seed sources appeared to have very similar temperature requirements, with night optima about the same of slightly lower than daylight optima.

Tree provenance is important in artificial regeneration. Good provenance takes into account suitable tree genetics and a good environmental fit for planted / seeded trees in a forest stand. The wrong genotype can lead to failed regeneration, or poor trees that are prone to pathogens and undesired outcomes.

Artificial regeneration has been a more common method involving planting because it is more dependable than natural regeneration. Planting can involve using seedlings (from a nursery), (un)rooted cuttings, or seeds.

Whichever method is chosen it can be assisted by tending techniques also known as intermediate stand treatments.

The fundamental genetic consideration in artificial regeneration is that seed and planting stock must be adapted to the planting environment. Most commonly, the method of managing seed and stock deployment is through a system of defined seed zones, within which seed and stock can be moved without risk of climatic maladaptation. Ontario adopted a seed zone system in the 1970s based on Hills' (1952) site regions and provincial resource district boundaries, but Ontario's seed zones are now based on homogeneous climatic regions developed with the Ontario Climate Model. The regulations stipulate that source-identified seedlots may be either a general collection, when only the seed zone of origin is known, or a stand collection from a specific latitude and longitude. The movement of general-collection seed and stock across seed zone boundaries is prohibited, but the use of stand-collection seed and stock in another seed zone is acceptable when the Ontario Climate Model shows that the planting site and place of seed origin are climatically similar. The 12 seed zones for white spruce in Quebec are based mainly on ecological regions, with a few modifications for administrative convenience.

Seed quality varies with source. Seed orchards produce seed of the highest quality, then, in order of decreasing seed quality produced, seed production areas and seed collection areas follow, with controlled general collections and uncontrolled general collections producing the least characterized seed.

Seeds

Dewinging, Extraction

When seed is first separated from cones it is mixed with foreign matter, often 2 to 5 times the volume of the seed. The more or less firmly attached membranous wings on the seed must be detached before it is cleaned of foreign matter. The testa must not incur damage during the dewinging process. Two methods have been used, dry and wet. Dry seed may be rubbed gently through a sieve that has a mesh through which only seed without wings can pass. Large quantities of seed can be processed in dewinging machines, which use cylinders of heavy wire mesh and rapidly revolving stiff brushes within to remove the wings. In the wet process, seed with wings attached are spread out 10 cm to 15 cm deep on a tight floor and slightly moistened throughout; light leather flails are used to free seed from the wings. Wang (1973) described a unique wet dewinging procedure using a cement mixer, used at the Petawawa tree seed processing facility. Wings of white and Norway spruce seed can be removed by dampening the seed slightly before it is run through a fanning mill for the last time. Any moistened seed must be dried before fermentation or moulding sets in.

Seed Viability

A fluorescein diacetate (FDA) biochemical viability test for several species of conifer seed, including white spruce, estimates the proportion of live seed (viability) in a seedlot, and hence the percentage germination of a seedlot. The accuracy of predicting percentage germination was within +/- 5 for most seedlots. White spruce seed can be tested for viability by an indirect method, such as the fluorescein diacetate (FDA) test or 'Ultra-sound'; or by the direct growth method of 'germination'. Samples of white spruce seed inspected by Toumey and Stevens (1928) varied in viability from 50% to 100%, but averaged 93%. Rafn (1915) reported 97% viability for white spruce seed.

Germinative Testing

The results of a germination test are commonly expressed as *germinative capacity* or a *germination percentage*, which is the percentage of seeds that germinate during a period of time, ending when germination is practically complete. During extraction and processing, white spruce seeds gradually lost moisture, and total germination increased. Mittal et al. (1987) reported that white spruce seed from Algonquin Park, Ontario, obtained the maximum rate (94% in 6 days) and 99% total germination in 21 days after 14-week pre-chilling. The pre-treatment of 1% sodium hypochlorite increased germinability.

Encouraged by Russian success in using ultrasonic waves to improve the germinative energy and percentage germination of seeds of agricultural crops, Timonin (1966) demonstrated benefits to white spruce germination after exposure of seeds to 1, 2, or 4 minutes of ultrasound generated by an M.S.E. ultrasonic disintegrator with a power consumption of 280 VA and power impact of 1.35 amperes. However, no seeds germinated after 6 minutes of exposure to ultrasound.

Seed Dormancy

Seed dormancy is a complex phenomenon and is not always consistent within species. Cold stratification of white spruce seed to break dormancy has been specified as a requirement, but Heit (1961) and Hellum (1968) regarded stratification as unnecessary. Cone handling and storage conditions affect dormancy in that cold, humid storage (5 °C, 75% to 95% relative humidity) of the cones prior to extraction seemingly eliminated dormancy by overcoming the need to stratify. Periods of cold, damp weather during the period of cone storage might provide natural cold (stratification) treatment. Once dormancy was removed in cone storage, subsequent kiln-drying and seed storage did not reactivate dormancy.

Haddon and Winston (1982) found a reduction in viability of stratified seeds after 2 years of storage and suggested that stress might have been caused by stratification, e.g., by changes in seed biochemistry, reduced embryo vigor, seed aging or actual damage to the embryo. They further questioned the quality of the 2-year-old seed even though high germination occurred in the samples that were not stratified.

Cold Stratification

Cold stratification is the term applied to the storing of seeds in (and, strictly, in layers with) a moist medium, often peat or sand, with a view to maintaining viability and overcoming dormancy. Cold stratification is the term applied to storage at near-freezing temperatures, even if no medium is used. A common method of cold stratification, is to soak seed in tap water for up to 24 h, superficially dry it, then store moist for some weeks or even months at temperatures just above freezing. Although Hellum (1968) found that cold stratification of an Alberta seed source led to irregular germination, with decreasing germination with increasing length of the stratification period, Hocking's (1972) paired test with stratified and nonstratified Alberta seed from several sources revealed no trends in response to stratification. Hocking suggested that seed maturity, handling, and storage needed to be controlled before the need for stratification could be determined. Later, Winston and Haddon (1981) found that the storage of white spruce cones for 4 weeks at 5 °C prior to extraction obviated the need for stratification.

Seed Ripeness

Seed maturity cannot be predicted accurately from cone flotation, cone moisture content, cone specific gravity; but the province of B.C. found embryo occupying 90% + of the corrosion cavity and megagametophyte being firm and whitish in colour are the best predictors for white spruce in B.C., and Quebec can forecast seed maturity some weeks in advance by monitoring seed development in relation to heat-sums and the phenological progression of the inflorescence of fireweed (*Epilobium angustifolium* L.), an associated plant species. Cone collection earlier than one week before seed maturity would reduce seed germination and viability during storage. Four stages of maturation were determined by monitoring carbohydrates, polyols, organic acids, respiration, and metabolic activity. White spruce seeds require a 6-week post-harvest ripening period in the cones to obtain maximum germinability, however, based on cumulative degree-days, seed from the same trees and stand showed 2-week cone storage was sufficient.

Forest Tree Plantations

Plantation Establishment Criteria

Plantations may be considered successful when outplant performance satisfies certain criteria. The term "free growing" is applied in some jurisdictions. Ontario's "Free-to-Grow" (FTG) equivalent relates to a forest stand that meets a minimum stocking standard and height requirement, and is essentially free of competition from surrounding vegetation that might impede growth. The FTG concept was introduced with the advent of the Forest Management Agreement program in Ontario in 1980 and became applicable to all management units in 1986. Policy, procedures, and methodologies readily applicable by forest unit managers to assess the effectiveness of regeneration programs were still under development during the Class Environmental Assessment hearings.

In British Columbia, the Forest Practices Code (1995) governs performance criteria. To minimize the subjectivity of assessing deciduous competition as to whether or not a plantation is established, minimum specifications of number, health, height, and competition have been specified in British Columbia. However, minimum specifications are still subjectively set and may need to be fine-tuned in order to avoid unwarranted delay in according established status to a plantation. For example, a vigorous white spruce with a strong, multi-budded leading shoot and its crown fully exposed to light on 3 sides would not qualify as free-growing in the current British Columbia Code but would hardly warrant description as unestablished.

Competition

Competition arises when individual organisms are sufficiently close together to incur growth constraint through mutual modification of the local environment. Plants may compete for light, moisture and nutrients, but seldom for space *per se*. Vegetation management directs more of the site's resources into usable forest products, rather than just eliminating all competing plants. Ideally, site preparation ameliorates competition to levels that relieve the outplant of constraints severe enough to cause prolonged check.

The diversity of boreal and sub-boreal broadleaf-conifer mixed tree species stands, commonly referred to as the "mixedwoods", largely preclude the utility of generalizations and call for the development of management practices incorporating the greater inherent complexity of broadleaf-conifer mixtures, relative to single-species or mixed-species conifer forest. After harvesting or other disturbance, mixedwood stands commonly enter a prolonged period in which hardwoods overtop the coniferous component, subjecting them to intense competition in an understorey. It is well established that the regeneration and growth potential of understorey conifers in mixedwood stands is correlated to the density of competing hardwoods. To help apply "free-to-grow" regulations in British Columbia and Alberta, management guidelines based on distance-dependent relations within a limited radius of crop trees were developed, but Lieffers et al. (2002) found that free-growing stocking standards did not adequately characterize light competition between broadleaf and conifer components in boreal mixedwood stands, and further noted that adequate sampling using current approaches would be operationally prohibitive.

Many promising plantations have failed through lack of tending. Young crop trees are often ill-equipped to fight it out with competition resurgent following initial site preparation and planting.

Perhaps the most direct evaluation of the effect of competition on plantation establishment is provided by an effective herbicide treatment. The fact that herbicide treatment does not always produce positive results should not obscure the demonstrated potential of herbicides for significantly promoting plantation establishment. Factors that can vitiate the effectiveness of a herbicide treatment include: weather, especially temperature, prior to and during application; weather, especially wind, during application; weather, especially precipitation, in the 12 to 24 hours after application; vegetation characteristics, including species, size, shape, phenological stage, vigour, and distribution of weeds; crop characteristics, including species, phenology, and condition; the effects of other treatments, such as preliminary shearblading, burning or other prescribed or accidental site preparation; and the herbicide used, including dosage, formulation, carrier, spreader, and mode of application. There is a lot that can go wrong, but a herbicide treatment can be as good or better than any other method of site preparation.

Competition Indices

The study of competition dynamics requires both a measure of the competition level and a measure of crop response. Various competition indices have been developed, e.g., by Bella (1971) and Hegyi (1974) based on stem diameter, by Arney (1972), Ek and Monserud (1974), and Howard and Newton (1984) based on canopy development, and Daniels (1976), Wagner (1982), and Weiner (1984) with proximity-based models. Studies generally considered tree response to competition in terms of absolute height or basal area, but Zedaker (1982) and Brand (1986) sought to quantify crop tree size and environmental influences by using relative growth measures.

Tending

Tending is the term applied to pre-harvest silvicultural treatment of forest crop trees at any stage after initial planting or seeding. The treatment can be of the crop itself (e.g., spacing, pruning, thinning, and improvement cutting) or of competing vegetation (e.g., weeding, cleaning).

Planting

How many trees per unit area (spacing) that should be planted is not an easily answered question. Establishment density targets or regeneration standards have commonly been based on traditional practice, with the implicit aim of getting the stand quickly to the free-to-grow stage. Money is wasted if more trees are planted than are needed to achieve desired stocking rates, and the chance to establish other plantations is proportionately diminished. Ingress (natural regeneration) on a site is difficult to predict and often becomes surprisingly evident only some years after planting has been carried out. Early stand development after harvesting or other disturbance undoubtedly varies greatly among sites, each of which has its own peculiar characteristics.

For all practical purposes, the total volume produced by a stand on a given site is constant and optimum for a wide range of density or stocking. It can be decreased, but not increased, by altering the amount of growing stock to levels outside this range. Initial density affects stand development in that close spacing leads to full site utilization more quickly than wider spacing. Economic operability can be advanced by wide spacing even if total production is less than in closely spaced stands.

Beyond the establishment stage, the relationship of average tree size and stand density is very important. Various density-management diagrams conceptualizing the density-driven stand dynamics have been developed. Smith and Brand's (1988) diagram has mean tree volume on the vertical axis and the number of trees/ha on the horizontal axis: a stand can either have many little trees or a few big ones. The self-thinning line shows the largest number of trees of a given size/ha that can be carried at any given time. However, Willcocks and Bell (1995) caution against using such diagrams unless specific knowledge of the stand trajectory is known.

In the Lake States, plantations have been made with the spacing between trees varying from 3 by 3 to 10 by 10 feet (0.9 m by 0.9 m to 3.0 m by 3.0 m). Kittredge recommended that no fewer than 600 established trees per acre (1483/ha) be present during the early life of a plantation. To insure this, at least 800 trees per acre (1077/ha) should be planted where 85% survival may be expected, and at least 1200/ac (2970/ha) if only half of them can be expected to live. This translates into recommended spacings of 5 by 5 to 8 by 8 feet (1.5 m by 1.5 m to 2.4 m by 2.4 m) for plantings of conifers, including white spruce in the Lake States.

Enrichment Planting

A strategy for enhancing natural forests' economic value is to increase their concentration of economically important, indigenous tree species by planting seeds or seedlings for future harvest, which can be accomplished with enrichment planting (EP). This means increasing the planting density (i.e., the numbers of plants per hectare) in an already growing forest stand."

Release Treatments

- Weeding: A process of getting rid of saplings' or seedlings' competition by mowing, application of herbicide, or other method of removal from the surroundings.
- Cleaning: Release of select saplings from competition by overtopping trees of a comparable age. The treatment favors trees of a desired species and stem quality.
- Liberation cutting: A treatment that releases tree seedling or saplings by removing older overtopping trees.

Spacing

Over-crowded regeneration tends to stagnate. The problem is aggravated in species that have little self-pruning ability, such as white spruce. Spacing is a thinning (of natural regeneration), in which all trees other than those selected for retention at fixed intervals are cut. The term juvenile spacing is used when most or all of the cut trees are unmerchantable. Spacing can be used to obtain any of a wide range of forest management objectives, but it is especially undertaken to reduce density and control stocking in young stands and prevent stagnation, and to shorten the rotation, i.e., to speed the production of trees of a given size. Volume growth of individual trees and the merchantable growth of stands are increased. The primary rationale for spacing is that thinning is the projected decline in maximum allowable cut. And since wood will be concentrated on fewer, larger, and more uniform stems, operating and milling costs will be minimized.

Methods for spacing may be: manual, using various tools, including power saws, brush saws, and clippers; mechanical, using choppersand mulchers; chemical; or combinations of several methods. One treatment has had notable success in spacing massively overstocked (<100 000 stems/ha) natural regeneration of spruce and fir in Maine. Fitted to helicopter, the Thru-Valve boom emits herbicide spray droplets 1000 µm to 2000 µm in diameter at very low pressure. Swaths 1.2 m wide and leave strips 2.4 m wide were obtained with "knife-edge" precision when the herbicide was applied by helicopter flying at a height of 21 m at a speed of 40–48 km/h. It seems likely that no other method could be as cost-effective.

Twenty years after spacing to 2.5 × 2.5 m, 30-year-old mixed stands of balsam fir and white spruce in the Green River watershed, New Brunswick, averaged 156.9 m^3/ha.

A spacing study of 3 conifers (white spruce, red pine and jack pine) was established at Moodie, Manitoba, on flat, sandy, nutritionally poor soils with a fresh moisture regime. Twenty years after planting, red pine had the largest average dbh, 15% greater than jack pine, while white spruce dbh was less than half that of the pines. Crown width showed a gradual increase with spacing for all 3 conifers. Results to date were suggesting optimum spacings between 1.8 m and 2.4 m for both pines; white spruce was not recommended for planting on such sites.

Comparable data are generated by espacement trials, in which trees are planted at a range of densities. Spacings of 1.25 m, 1.50 m, 1.75 m, 2.00 m, 2.50 m, and 3.00 m on 4 site classes were used in the 1922 trial at Petawawa, Ontario. In the first of 34 old field white spruce plantations used to investigate stand development in relation to spacing at Petawawa, Ontario, regular rows were planted at average spacings of from 4 × 4 to 7 × 7 feet (1.22 m × 1.22 m to 2.13 m × 2.13 m). Spacings up to 10 × 10 feet (3.05 m × 3.03 m) were subsequently included in the study. Yield tables based on 50 years of data showed:

a) Except for merchantable volumes at age 20 and site classes 50 and 60, closer spacings gave greater standing volumes at all ages than did wider spacings, the relative difference decreasing with age.

b) Merchantable volume as a proportion of total volume increases with age, and is greater at wider than at closer spacings.

c) Current annual volume increment culminates sooner at closer than at wider spacings.

A smaller espacement trial, begun in 1951 near Thunder Bay, Ontario, included white spruce at spacings of 1.8 m, 2.7 m, and 3.6 m. At the closest spacing, mortality had begun at 37 years, but not at the wider spacings.

The oldest interior spruce espacement trial in British Columbia was established in 1959 near Houston in the Prince Rupert Forest Region. Spacings of 1.2 m, 2.7 m, 3.7 m, and 4.9 m were used, and trees were measured 6, 12, 16, 26, and 30 years after planting. At wide espacements, trees developed larger diameters, crowns, and branches, but (at 30 years) basal area and total volume/ha were greatest in the closest espacement (Table 6.38). In more recent trials in the Prince George Region of British Columbia (Table 6.39) and in Manitoba, planting density of white spruce had no effect on growth after up to 16 growing seasons, even at spacings as low as 1.2 m. The slowness of juvenile growth and of crown closure delay the response to intra-competition. Initially, close spacing might even provide a positive nurse effect to offset any negative response to competition.

Thinning

Thinning is an operation that artificially reduces the number of trees growing in a stand with the aim of hastening the development of the remainder. The goal of thinning is to control the amount and distribution of available growing space. By altering stand density, foresters can influence the growth, quality, and health of residual trees. It also provides an opportunity to capture mortality and cull the commercially less desirable, usually smaller and malformed, trees. Unlike regeneration treatments, thinnings are not intended to establish a new tree crop or create permanent canopy openings.

Thinning greatly influences the ecology and micro-meteorology of the stand, lowering the inter-tree competition for water. The removal of any tree from a stand has repercussions on the remaining trees both above-ground and below. Silvicultural thinning is a powerful tool that can be used to influence stand development, stand stability, and the characteristics of the harvestable products.

When considering intensive conifer plantations designed for maximum production, it is essential to remember that tending and thinning regimes and wind and snow damage are intimately related.

Previous studies have demonstrated that repeated thinnings over the course of a forest rotation increase carbon stores relative to stands that are clear-cut on short rotations and that the carbon benefits differ according to thinning method (e.g., thinning from above versus below).

Precommercial Thinning

When natural regeneration or artificial seeding has resulted in dense, overstocked young stands, natural thinning will in most cases eventually reduce stocking to more silviculturally desirable levels. But by the time some trees reach merchantable size, others will be overmature and defective, and others will still be unmerchantable. The yield of merchantable wood can be greatly increased and the rotation shortened by precommercial thinning. Mechanical and chemical methods have been applied, but their costliness has militated against their ready adoption.

Pruning

Pruning, as a silvicultural practice, refers to the removal of the lower branches of the young trees (also giving the shape to the tree) so clear knot-free wood can subsequently grow over the branch stubs. Clear knot-free lumber has a higher value. Pruning has been extensively carried out in the Radiata pine plantations of New Zealand and Chile, however the development of Finger joint technology in the production of lumber and mouldings has led to many forestry companies reconsidering their pruning practices. "Brashing" is an alternative name for the same process. Pruning can be done to all trees, or more cost effectively to a limited number of trees. There are two types of pruning: natural or self-pruning and artificial pruning. Most cases of self-pruning happen when branches do not receive enough sunlight and die. Wind can also take part in natural pruning which can break branches. Artificial pruning is where people are paid to come and cut the branches. Or it can be natural, where trees are planted close enough that the effect is to cause self-pruning of low branches as energy is put into growing up for light reasons and not branchiness.

Stand Conversion

The term *stand conversion* refers to a change from one silvicultural system to another and includes *species conversion*, i.e., a change from one species (or set of species) to another. Such change can be effected intentionally by various silvicultural means, or incidentally by default e.g., when high-grading has removed the coniferous content from a mixedwood stand, which then becomes exclusively self-perpetuating aspen. In general, such sites as these are the most likely to be considered for conversion.

Growth and Yield

In discussing yields that might be expected from the Canadian spruce forests, Haddock (1961) noted that Wright's (1959) quotation of spruce yields in the British Isles of 220 cubic feet per acre (15.4 m^3/ha) per year and in Germany of 175 cubic feet per acre (12.25 m^3/ha) per year was misleading, at least if it was meant to imply that such yields might be approached in the Boreal Forest Region of Canada. Haddock thought that Wright's suggestion of 20 to 40 (average 30) cubic feet per acre (1.4 m^3/ha to 2.8 m^3/ha (average 2.1 m^3/ha) per year was more reasonable, but still somewhat optimistic.

The principal way forest resource managers influence growth and yield is to manipulate the mixture of species and number (density) and distribution (stocking) of individuals that form the canopy of the stand. Species composition of much of the boreal forest in North America already differs greatly from its pre-exploitation state. There is less spruce and more hardwoods in the second-growth forest than in the original forest; Hearnden et al. (1996) calculated that the spruce cover type had declined from 18% to only 4% of the total forested area in Ontario. Mixedwood occupies a greater proportion of Ontario's second-growth forest (41%) than in the original (36%), but its component of white spruce is certainly much diminished.

Growth performance is certainly influenced by site conditions and thus by the kind and degree of site preparation in relation to the nature of the site. It is important to avoid the assumption that site preparation of a particular designation will have a particular silvicultural outcome. Scarification, for instance, not only covers a wide range of operations that scarify, but also any given way of scarifying can have significantly different results depending on site conditions at the time of treatment. In point of fact, the term is commonly misapplied. *Scarification* is defined as "Loosening the top soil of open areas, or breaking up the forest floor, in preparation for regenerating by direct seeding or natural seedfall", but the term is often misapplied to practices that include scalping, screefing, and blading, which pare off low and surface vegetation, together with most off its roots to expose a weed-free surface, generally in preparation for sowing or planting thereon.

Thus, it is not surprising that literature can be used to support the view that the growth of seedlings on scarified sites is much superior to that of growth on similar sites that have not been scarified, while other evidence supports the contrary view that scarification can reduce growth. Detrimental results can be expected from scarification that impoverishes the rooting zone or exacerbates edaphic or climatic constraints.

Burning site preparation has enhanced spruce seedling growth, but it must be supposed that burning could be detrimental if the nutrient capital is significantly depleted.

An obvious factor greatly influencing regeneration is competition from other vegetation. In a pure stand of Norway spruce, for instance, Roussel (1948) found the following relationships:

Percent cover (%)	Vegetation Description
Below 1	No vegetation
1-3	Moss carpet with a few fir seedlings
4-10	Herbaceous plants appear
10-25	Bramble, herbs, fairly vigorous spruce seedlings
>25	Herbs, brambles very dense, vigorous, no moss

A factor of some importance in solar radiation–reproduction relationships is excess heating of the soil surface by radiation. This is especially important for seedlings, such as spruce, whose first leaves do not shade the base of the stem at the soil surface. Surface temperatures in sandy soils on occasion reach lethal temperatures of 50 °C to 60 °C.

Common Methods of Harvesting

Silvicultural regeneration methods combine both the harvest of the timber on the stand and re-establishment of the forest. The proper practice of sustainable forestry should mitigate the potential negative impacts, but all harvest methods will have some impacts on the land and residual stand. The practice of sustainable forestry limits the impacts such that the values of the forest are maintained in perpetuity. Silvicultural prescriptions are specific solutions to a specific set of circumstances and management objectives. Following are some common methods:

Clearcut Harvesting

Conventional clearcut harvesting is relatively simple: all trees on a cutblock are felled and bunched with bunches aligned to the skidding direction, and a skidder then drags the bunches to the closest log deck. Feller-buncher operators concentrate on the width of the felled swath, the number of trees in a bunch, and the alignment of the bunch. Providing a perimeter boundary is felled during daylight, night-shift operations can continue without the danger of trespassing beyond the block. Productivity of equipment is maximized because units can work independently of one another.

Clearcutting

An even-aged regeneration method that can employ either natural or artificial regeneration. It involves the complete removal of the forest stand at one time. Clearcutting can be biologically appropriate with species that typically regenerate from stand replacing fires or other major disturbances, such as Lodgepole Pine (*Pinus contorta*). Alternatively, clearcutting can change the dominating species on a stand with the introduction of non-native and invasive species as was shown at the Blodgett Experimental Forest near Georgetown California. Additionally, clearcutting can prolong slash decomposition, expose soil to erosion, impact visual appeal of a landscape and remove essential wildlife habitat. It is particularly useful in regeneration of tree species such as Douglas-fir (*Pseudotsuga menziesii*) which is shade intolerant. In addition, the general public's distaste for even-aged silviculture, particularly clearcutting, is likely to result in a greater role for uneven-aged management on public lands as well. Across Europe, and in parts of North America, even-aged, production-orientated and intensively managed plantations are beginning to be regarded in the same way as old industrial complexes: something to abolish or convert to something else.

Clearcutting will impact many site factors important in their effect on regeneration, including air and soil temperatures. Kubin and Kemppainen (1991), for instance, measured temperatures in northern Finland from 1974 through 1985 in 3 clear-felled areas and in 3 neighouring forest stands dominated by Norway spruce. Clear felling had no significant influence on air temperature at 2 m above the ground surface, but the daily air temperature maxima at 10 cm were greater in the clear-felled area than in the uncut forest, while the daily minima at 10 cm were lower. Night frosts were more common in the clear-felled area. Daily soil temperatures at 5 cm depth were 2 °C to 3 °C greater in the clear-felled area than in the uncut forest, and temperatures at depths of 50 cm and 100 cm were 3 °C to 5 °C greater. The differences between the clear-felled and uncut areas did not diminish during the 12 years following cutting.

Coppicing

A regeneration method which depends on the sprouting of cut trees. Most hardwoods, the coast redwood, and certain pines naturally sprout from stumps and can be managed through coppicing. Coppicing is generally used to produce fuelwood, pulpwood, and other products dependent on small trees. A close relative of coppicing is pollarding. Three systems of coppice woodland management are generally recognized: simple coppice, coppice with standards, and the coppice selection system.

In Compound coppicing or coppicing with standards, some of the highest quality trees are retained for multiple rotations in order to obtain larger trees for different purposes.

Direct Seeding

Prochnau (1963), 4 years after sowing, found that 14% of viable white spruce seed sown on mineral soil had produced surviving seedlings, at a seed:seedling ratio of 7.1:1. With Engelmann spruce, Smith and Clark (1960) obtained average 7th year seed:seedling ratios of 21:1 on scarified seedbeds on dry sites, 38:1 on moist sites, and 111:1 on litter seedbeds.

Group Selection

The group selection method is an uneven-aged regeneration method that can be used when mid-tolerant species regeneration is desired. The group selection method can still result in residual stand damage in dense stands, however directional falling can minimize the damage. Additionally, foresters can select across the range of diameter classes in the stand and maintain a mosaic of age and diameter classes.

Méthode Du Contrôle

Classical European silviculture achieved impressive results with systems such as Henri Biolley's *méthode du contrôle* in Switzerland, in which the number and size of trees harvested were determined by reference to data collected from every tree in every stand measured every 7 years.

While not designed to be applied to boreal mixedwoods, the *méthode du contrôle* is described briefly here to illustrate the degree of sophistication applied by some European foresters to the management of their forests. Development of management techniques that allowed for stand development to be monitored and guided into sustainable paths were in part a response to past experience, particularly in Central European countries, of the negative effects of pure, uniform stands with species often unsuited to the site, which greatly increased the risk of soil degradation and biotic diseases. Increased mortality and decreased increment generated widespread concern, especially after reinforcement by other environmental stresses.

More or less uneven-aged, mixed forests of preponderantly native species, on the other hand, treated along natural lines, have proved to be healthier and more resistant to all kinds of external dangers; and in the long run such stands are more productive and easier to protect.

However, irregular stands of this type are definitely more difficult to manage—new methods and techniques had to be sought particularly for the establishment of inventories, as well as increment control and yield regulation. In Germany, for instance, since the beginning of the nineteenth century under the influence of G.L. Hartig (1764–1837), yield regulation has been effected almost exclusively by allotment or formula methods based on the conception of the uniform normal forest with a regular succession of cutting areas.

In France, on the other hand, efforts were made to apply another kind of forest management, one that aimed to bring all parts of the forest to a state of highest productive capacity in perpetuity. In 1878, the French forester A. Gurnaud (1825–1898) published a description of a *méthode du contrôle* for determining increment and yield. The method was based on the fact that through careful, selective harvesting, the productivity of the residual stand can be improved, because timber is removed as a cultural operation. In this method, the increment of stands is accurately determined periodically with the object of gradually converting the forest, through selective management and continuous experimentation, to a condition of equilibrium at maximum productive capacity.

Henri Biolley (1858–1939) was the first to apply Gurnaud's inspired ideas to practical forestry. From 1890 on, he managed the forests of his Swiss district according to these principles, devoting himself for almost 50 years to the study of increment and a treatment of stands directed towards the highest production, and proving the practicability of the check method. In 1920, he published this study giving a theoretical basis of management of forests under the check method, describing the procedures to be applied in practice (which he partly developed and simplified), and evaluating the results.

Biolley's pioneering work formed the basis upon which most Swiss forest management practices were later developed, and his ideas have been generally accepted. Today, with the trend of intensifying forest management and productivity in most countries, the ideas and application of careful, continuous treatment of stands with the aid of the volume check method are meeting with ever-growing interest. In Britain and Ireland, for example, there is increased application of Continuous Cover Forestry principles to create permanently irregular structures in many woodlands.

Patch Cut

Row and Broadcast Seeding

Spot and row seeders use less seed that does broadcast ground or aerial seeding but may induce clumping. Row and spot seeding confer greater ability to control seed placement than does broadcast seeding. Also, only a small percentage of the total area needs to be treated.

In the aspen type of the Great Lakes region, direct sowing of the seed of conifers has usually failed. However, Gardner (1980) after trials in Yukon, which included broadcast seeding of white spruce seed at 2.24 kg/ha that secured 66.5% stocking in the Scarified Spring Broadcast treatment 3 years after seeding, concluded that the technique held "considerable promise".

Seed-tree

An even-aged regeneration method that retains widely spaced residual trees in order to provide uniform seed dispersal across a harvested area. In the seed-tree method, 2-12 seed trees per acre (5-30/ha) are left standing in order to regenerate the forest. They will be retained until regeneration has become established at which point they may be removed. It may not always be economically viable or biologically desirable to re-enter the stand to remove the remaining seed trees. Seed-tree cuts can also be viewed as a clearcut with natural regeneration and can also have all of the problems associated with clearcutting. This method is most suited for light-seeded species and those not prone to windthrow.

Selection Systems

Selection systems are appropriate where uneven stand structure is desired, particularly where the need to retain continuous cover forest for aesthetic or environmental reasons outweighs other management considerations. Selection logging has been suggested as being of greater utility than shelterwood systems in regenerating old-growth Engelmann Spruce Sub-alpine Fir (ESSF) stands in southern British Columbia. In most areas, selection logging favours regeneration of fir more than the more light-demanding spruce. In some areas, selection logging can be expected to favour spruce over less tolerant hardwood species (Zasada 1972) or lodgepole pine.

Shelter Spot Seeding

The use of shelters to improve germination and survival in spot seedings seeks to capture the benefits of greenhouse culture, albeit miniature. The Hakmet seed shelter, for instance, is a semi-transparent plastic cone 8 cm high, with openings of 7 cm diameter in the 7.5 cm diameter base and 17 mm diameter in the 24 mm diameter top. This miniature greenhouse increases air humidity, reduces soil desiccation, and raises air and soil temperatures to levels more favourable to germination and seedling growth than those offered by unprotected conditions. The shelter is designed to break down after a few years of exposure to ultraviolet radiation.

Seed shelters and spring sowing significantly improved stocking compared with bare spot seeding, but sheltering did not significantly improve growth. Stocking of bare seedspots was extremely low, possibly due to smothering of seedlings by abundant broadleaf and herbaceous litter, particularly that from aspen and red raspberry, and exacerbated by strong competition from graminoids and raspberry.

Cone shelters (Cerkon™) usually produced greater survival than unsheltered seeding on scarified seedspots in trials of direct seeding techniques in interior Alaska, and funnel shelters (Cerbel™) usually produced greater survival than unsheltered seeding on non-scarified seedspots. Both shelter types are manufactured by AB Cerbo in Trollhättan, Sweden. Both are made of light-degradable, white, opaque plastic, and are 8 cm high when installed.

White spruce seed was sown in Alaska on a burned site in summer 1984, and protected by white plastic cones on small spots scarified by hand, or by white funnels placed directly into the residual ash and organic material. A group of 6 ravens (*Corvus corax*) was observed in the area about 1 week after sowing was completed in mid-June. Damage averaged 68% with cones and 50% with funnels on an upland area, and 26% with funnels on a floodplain area. Damage by ravens was only 0.13% on unburned but otherwise similar areas.

In seeding trials in Manitoba between 1960 and 1966 aimed at converting aspen stands to spruce–aspen mixedwoods, 1961 scarification in the Duck Mountain Provincial Forest remained receptive to natural seeding for many years.

Shelterwood

In general terms, the shelterwood system is a series of partial cuts that removes the trees of an existing stand over several years and eventually culminates in a final cut that creates a new even-aged stand. It is an even-aged regeneration method that removes trees in a series of three harvests: 1) Preparatory cut; 2) Establishment cut; and 3) Removal cut. The success of practising a shelterwood system is closely related to: 1. the length of the regeneration period, i.e. the time from the shelterwood cutting to the date when a new generation of trees has been established; 2.the quality of the new tree stand with respect to stand density and growth; and 3.the value increment of the shelter trees. Information on the establishment, survival and growth of seedlings influenced by the cover of shelter trees, as well as on the growth of these trees, is needed as a basis for modelling the economic return of practising a shelterwood system. The method's objective is to establish new forest reproduction under the shelter of the retained trees. Unlike the seed-tree method, residual trees alter understory environmental conditions (i.e. sunlight, temperature, and moisture) that influence tree seedling growth. This method can also find a middle ground with the light ambiance by having less light accessible to competitors while still being able to provide enough light for tree regeneration. Hence, shelterwood methods are most often chosen for site types characterized by extreme conditions, in order to create a new tree generation within a reasonable time period. These conditions are valid foremost on level ground sites which are either dry and poor or moist and fertile.

Shelterwood Systems

Shelterwood systems involve 2, 3, or exceptionally more partial cuttings. A final cut is made once adequate natural regeneration has been obtained. The shelterwood system is most commonly applied as a 2-cut uniform shelterwood, first an initial regeneration (seed) cut, the second a final harvest cut. In stands less than 100 years old, a light preparatory cut can be useful. A series of intermediate cuts at intervals of 10–20 years has been recommended for intensively managed stands.

From operational or economic standpoints, however, there are disadvantages to the shelterwood system: harvesting costs are higher; trees left for deferred cutting may be damaged during the regeneration cut or related extraction operations; the increased risk of blowdown threatens the seed source; damage from bark beetles is likely to increase; regeneration may be damaged during the final cut and related extraction operations; the difficulty of any site preparation would be increased; and incidental damage to regeneration might be caused by any site preparation operations.

Single-tree Selection

The single-tree selection method is an uneven-aged regeneration method most suitable when shade tolerant species regeneration is desired. It is typical for older and diseased trees to be removed, thus thinning the stand and allowing for younger, healthy trees to grow. Single-tree selection can be very difficult to implement in dense or sensitive stands and residual stand damage can occur. This method is also disturbs the canopy layer the least out of all other methods.

Spot Seeding

Spot seeding was found to be the most economical and reliable of the direct seeding methods for converting aspen and paper birch to spruce and pine. In the Chippewa National Forest (Lake States), seed-spot sowing of 10 seeds each of white spruce and white pine under 40-year aspen after different degrees of cutting on gave second-season results clearly indicating the need to remove or disturb the forest floor to obtain germination of seeded white spruce and white pine.

Spot seeding of coniferous seed, including white spruce, has had occasional success, but several constraining factors commonly limit germination success: the drying out of the forest floor before the roots of germinants reach underlying moisture reserves; and, particularly under hardwoods, the smothering of small seedlings by snow-pressed leaf litter and lesser vegetation. Kittredge and Gervorkiantz (1929) determined that removal of the aspen forest floor increased germination percentage after the second season in seed spots of both white pine and white spruce, in 4 plots, from 2.5% to 5%, from 8% to 22%, from 1% to 9.5%, and from 0% to 15%.

Spot seeding requires less seed than broadcast seeding and tends to achieve more uniform spacing, albeit sometimes with clumping. The devices used in Ontario for manual spot seeding are the "oil can" seeder, seeding sticks, and shakers. The oil can is a container fitted with a long spout through which a predetermined number of seeds are released with each flick of the seeder.

Strip Cutting

Harvesting cutblocks where only a portion of the trees are to be removed is very different from clearcutting. First, trails must be located to provide access for the felling and skidding/forwarding equipment. These trails must be carefully located to ensure that the trees remaining meet the desired quality criteria and stocking density. Second, the equipment must not damage the residual stand. The further desiderata are outlined by Sauder (1995).

The dearth of seed and a deficiency of receptive seedbeds were recognized as major reasons for the lack of success of clearcut harvesting. One remedy attempted in British Columbia and Alberta has been alternate strip cutting. The greater seed source from uncut trees between the cut strips, and the disturbance to the forest floor within the cut strips could be expected to increase the amount of natural regeneration. Trees were cut to a diameter limit in the cut strips, but large trees in the leave strips often proved too much of a temptation and were cut too, thus removing those trees that would otherwise have been the major source of seed.

An unfortunate consequence of strip thinning was the build-up of spruce beetle populations. Shaded slash from the initial cut, together with an increase in the number of windthrown trees in the leave strips, provided conditions ideally suited to the beetle.

Underplanting

DeLong et al. (1991) suggested underplanting 30- to 40-year-old aspen stands, on the basis of the success of natural spruce in regenerating under stands of such stands: "By planting, spacing can be controlled enabling easier protection of the spruce during stand entry for harvesting of the aspen overstorey".

Variable Retention

A harvesting and regeneration method which is a relatively new silvicultural system that retains forest structural elements (stumps, logs, snags, trees, understory spieces and undisturbed layers of forest floor) for at least one rotation in order to preserve environmental values associated with structurally complex forests.

"Uneven-aged and even-aged methods differ in the scale and intensity of disturbance. Uneven-aged methods maintain a mix of tree sizes or ages within a habitat patch by periodically harvesting individual or small groups of trees, Even-aged methods harvest most or all of the overstory and create a fairly uniform habitat patch dominated by trees of the same age". Even-aged management systems have been the prime methods to use when studying the effects on birds.

Mortality

A survey in 1955–56 to determine survival, development, and the reasons for success or failure of conifer pulpwood plantations (mainly of white spruce) in Ontario and Quebec up to 32 years old found that the bulk of the mortality occurred within the first 4 years of planting, unfavourable site and climate being the main causes of failure.

Advance Growth

Naturally regenerated trees in an understorey prior to harvesting constitute a classic case of good news and bad news. Understorey white spruce is of particular importance in mixedwoods dominated by aspen, as in the B15, B18a, and B19a Sections of Manitoba, and elsewhere. Until the latter part of the last century, white spruce understorey was mostly viewed as money in the bank on a long-term, low interest deposit, with final yield to be realized after slow natural succession, but the resource became increasingly threatened with the intensification of harvesting of aspen. White spruce plantations on mixedwood sites proved expensive, risky, and generally unsuccessful. This prompted efforts to see what might be done about growing aspen and white spruce on the same landbase by protecting existing white spruce advance growth, leaving a range of viable crop trees during the first cut, then harvesting both hardwoods and spruce in the final cut. Information about the understorey component is critical to spruce management planning. The ability of then current harvesting technology and crews employed to provide adequate protection for white spruce understories was questioned by Brace and Bella. Specialized equipment and training, perhaps with financial incentives, may be needed to develop procedures that would confer the degree of protection needed for the system to be feasible. Effective understorey management planning requires more than improved mixedwood inventory.

Avoidance of damage to the understorey will always be a desideratum. Sauder's (1990) paper on mixedwood harvesting describes studies designed to evaluate methods of reducing non-trivial damage to understorey residuals that would compromise their chance of becoming a future crop tree. Sauder concluded that: (1) operational measures that protected residual stems may not unduly increase costs, (2) all felling, conifers and hardwoods, needs to be done in one operation to minimize the entry of the feller-buncher into the residual stand, (3) several operational procedures can reduce understorey damage, some of them without incurring extra costs, and (4) successful harvesting of treatment blocks depends primarily on the intelligent location of skid trails and landings. In summary, the key to protecting the white spruce understorey without sacrificing logging efficiency is a combination of good planning, good supervision, the use of appropriate equipment, and having conscientious, well-trained operators.Even the best plan will not reduce understorey damage unless its implementation is supervised.

New stands need to be established to provide for future supply of commercial white spruce from 150 000 ha of boreal mixedwoods in 4 of Rowe's (1972) regional Forest Sections straddling Alberta, Saskatchewan, and Manitoba, roughly from Peace River AB to Brandon MB. In the 1980s, with harvesting using conventional equipment and procedures, a dramatic increase in the demand for aspen posed a serious problem for the associated spruce understorey. Formerly, white spruce in the understories had developed to commercial size through natural succession under the protection of the hardwoods. Brace articulated a widespread concern: "The need for protection of spruce as a component of boreal mixedwoods goes beyond concern for the future commercial softwood timber supply. Concerns also include fisheries and wildlife habitat, aesthetics and recreation, a general dissatisfaction with cleacutting in mixedwoods and a strong interest in mixedwood perpetuation, as expressed recently in 41 public meetings on forestry development in northern Alberta...".

On the basis of tests of 3 logging systems in Alberta, Brace (1990) affirmed that significant amounts of understorey can be retained using any of those systems provided that sufficient effort is directed towards protection. Potential benefits would include increased short-term softwood timber supply, improved wildlife habitat and cutblock aesthetics, as well as reduced public criticism of previous logging practices. Stewart et al. (2001) developed statistical models to predict the natural establishment and height growth of understorey white spruce in the boreal mixedwood forest in Alberta using data from 148 permanent sample plots and supplementary information about height growth of white spruce regeneration and the amount and type of available substrate. A discriminant model correctly classified 73% of the sites as to presence or absence of a white spruce understorey, based on the amount of spruce basal area, rotten wood, ecological nutrient regime, soil clay fraction, and elevation, although it explained only 30% of the variation in the data. On sites with a white spruce understorey, a regression model related the abundance of regeneration to rotten wood cover, spruce basal area, pine basal area, soil clay fraction, and grass cover ($R^2 = 0.36$). About half the seedlings surveyed grew on rotten wood, and only 3% on mineral soil, and seedlings were 10 times more likely to have established on these substrates than on litter. Exposed mineral soil covered only 0.3% of the observed transect area.

Advance Growth Management

Advance growth management, i.e., the use of suppressed understorey trees, can reduce reforestation costs, shorten rotations, avoid denuding the site of trees, and also reduce adverse impacts on aesthetic, wildlife, and watershed values. To be of value, advance growth must have acceptable species composition and distribution, have potential for growth following release, and not be vulnerable to excessive damage from logging.

The age of advance growth is difficult to estimate from its size, as white that appears to be 2- to 3-year-old may well be more than 20 years old. However, age does not seem to determine the ability of advance growth of spruce to respond to release, and trees older than 100 years have shown rapid rates of growth after release. Nor is there a clear relationship between the size of advance growth and its growth rate when released.

Where advance growth consists of both spruce and fir, the latter is apt to respond to release more quickly than the former, whereas spruce does respond. If the ratio of fir to spruce is large, however, the greater responsiveness to release of fir may subject the spruce to competition severe enough to negate much of the effect of release treatment. Even temporary relief from shrub competition has increased height growth rates of white spruce in northwestern New Brunswick, enabling the spruce to overtop the shrubs.

Site Preparation

Site preparation is any of various treatments applied to a site in order to ready it for seeding or planting. The purpose is to facilitate the regeneration of that site by the chosen method. Site preparation may be designed to achieve, singly or in any combination: improved access, by reducing or rearranging slash, and amelioration of adverse forest floor, soil, vegetation, or other biotic factors. Site preparation is undertaken to ameliorate one or more constraints that would otherwise be likely to thwart the objectives of management. A valuable bibliography on the effects of soil temperature and site preparation on subalpine and boreal tree species has been prepared by McKinnon et al. (2002).

Site preparation is the work that is done before a forest area is regenerated. Some types of site preparation are burning.

Burning

Broadcast burning is commonly used to prepare clearcut sites for planting, e.g., in central British Columbia, and in the temperate region of North America generally.

Prescribed burning is carried out primarily for slash hazard reduction and to improve site conditions for regeneration; all or some of the following benefits may accrue:

a) Reduction of logging slash, plant competition, and humus prior to direct seeding, planting, scarifying or in anticipation of natural seeding in partially cut stands or in connection with seed-tree systems.

b) Reduction or elimination of unwanted forest cover prior to planting or seeding, or prior to preliminary scarification thereto.

c) Reduction of humus on cold, moist sites to favour regeneration.

d) Reduction or elimination of slash, grass, or brush fuels from strategic areas around forested land to reduce the chances of damage by wildfire.

Prescribed burning for preparing sites for direct seeding was tried on a few occasions in Ontario, but none of the burns was hot enough to produce a seedbed that was adequate without supplementary mechanical site preparation.

Changes in soil chemical properties associated with burning include significantly increased pH, which Macadam (1987) in the Sub-boreal Spruce Zone of central British Columbia found persisting more than a year after the burn. Average fuel consumption was 20 to 24 t/ha and the forest floor depth was reduced by 28% to 36%. The increases correlated well with the amounts of slash (both total and ≥7 cm diameter) consumed. The change in pH depends on the severity of the burn and the amount consumed; the increase can be as much as 2 units, a 100-fold change. Deficiencies of copper and iron in the foliage of white spruce on burned clearcuts in central British Columbia might be attributable to elevated pH levels.

Even a broadcast slash fire in a clearcut does not give a uniform burn over the whole area. Tarrant (1954), for instance, found only 4% of a 140-ha slash burn had burned severely, 47% had burned lightly, and 49% was unburned. Burning after windrowing obviously accentuates the subsequent heterogeneity.

Marked increases in exchangeable calcium also correlated with the amount of slash at least 7 cm in diameter consumed. Phosphorus availability also increased, both in the forest floor and in the 0 cm to 15 cm mineral soil layer, and the increase was still evident, albeit somewhat diminished, 21 months after burning. However, in another study in the same Sub-boreal Spruce Zone found that although it increased immediately after the burn, phosphorus availability had dropped to below pre-burn levels within 9 months.

Nitrogen will be lost from the site by burning, though concentrations in remaining forest floor were found by Macadam (1987) to have increased in 2 of 6 plots, the others showing decreases. Nutrient losses may be outweighed, at least in the short term, by improved soil microclimate through the reduced thickness of forest floor where low soil temperatures are a limiting factor.

The *Picea/Abies* forests of the Alberta foothills are often characterized by deep accumulations of organic matter on the soil surface and cold soil temperatures, both of which make reforestation difficult and result in a general deterioration in site productivity; Endean and Johnstone (1974) describe experiments to test prescribed burning as a means of seedbed preparation and site amelioration on representative clear-felled *Picea/Abies* areas. Results showed that, in general, prescribed burning did not reduce organic layers satisfactorily, nor did it increase soil temperature, on the sites tested. Increases in seedling establishment, survival, and growth on the burned sites were probably the result of slight reductions in the depth of the organic layer, minor increases in soil temperature, and marked improvements in the efficiency of the planting crews. Results also suggested that the process of site deterioration has not been reversed by the burning treatments applied.

Ameliorative Intervention

Slash weight (the oven-dry weight of the entire crown and that portion of the stem < 4 inches in diameter) and size distribution are major factors influencing the forest fire hazard on harvested sites. Forest managers interested in the application of prescribed burning for hazard reduction and silviculture, were shown a method for quantifying the slash load by Kiil (1968). In west-central Alberta, he felled, measured, and weighed 60 white spruce, graphed (a) slash weight per merchantable unit volume against diameter at breast height (dbh), and (b) weight of fine slash (<1.27 cm) also against dbh, and produced a table of slash weight and size distribution on one acre of a hypothetical stand of white spruce. When the diameter distribution of a stand is unknown, an estimate of slash weight and size distribution can be obtained from average stand diameter, number of trees per unit area, and merchantable cubic foot volume. The sample trees in Kiil's study had full symmetrical crowns. Densely growing trees with short and often irregular crowns would probably be overestimated; open-grown trees with long crowns would probably be underestimated.

The need to provide shade for young outplants of Engelmann spruce in the high Rocky Mountains is emphasized by the U.S. Forest Service. Acceptable planting spots are defined as microsites on the north and east sides of down logs, stumps, or slash, and lying in the shadow cast by such material. Where the objectives of management specify more uniform spacing, or higher densities, than obtainable from an existing distribution of shade-providing material, redistribution or importing of such material has been undertaken.

Access

Site preparation on some sites might be done simply to facilitate access by planters, or to improve access and increase the number or distribution of microsites suitable for planting or seeding.

Wang et al. (2000) determined field performance of white and black spruces 8 and 9 years after outplanting on boreal mixedwood sites following site preparation (Donaren disc trenching versus no trenching) in 2 plantation types (open versus sheltered) in southeastern Manitoba. Donaren trenching slightly reduced the mortality of black spruce but significantly increased the mortality of white spruce. Significant difference in height was found between open and sheltered plantations for black spruce but not for white spruce, and root collar diameter in sheltered plantations was significantly larger than in open plantations for black spruce but not for white spruce. Black spruce open plantation had significantly smaller volume (97 cm^3) compared with black spruce sheltered (210 cm^3), as well as white spruce open (175 cm^3) and sheltered (229 cm^3) plantations. White spruce open plantations also had smaller volume than white spruce sheltered plantations. For transplant stock, strip plantations had a significantly higher volume (329 cm^3) than open plantations (204 cm^3). Wang et al. (2000) recommended that sheltered plantation site preparation should be used.

Mechanical

Up to 1970, no "sophisticated" site preparation equipment had become operational in Ontario, but the need for more efficacious and versatile equipment was increasingly recognized. By this time, improvements were being made to equipment originally developed by field staff, and field testing of equipment from other sources was increasing.

According to J. Hall (1970), in Ontario at least, the most widely used site preparation technique was post-harvest mechanical scarification by equipment front-mounted on a bulldozer (blade, rake, V-plow, or teeth), or dragged behind a tractor (Imsett or S.F.I. scarifier, or rolling chopper). Drag type units designed and constructed by Ontario's Department of Lands and Forests used anchor chain or tractor pads separately or in combination, or were finned steel drums or barrels of various sizes and used in sets alone or combined with tractor pad or anchor chain units.

J. Hall's (1970) report on the state of site preparation in Ontario noted that blades and rakes were found to be well suited to post-cut scarification in tolerant hardwood stands for natural regeneration of yellow birch. Plows were most effective for treating dense brush prior to planting, often in conjunction with a planting machine. Scarifying teeth, e.g., Young's teeth, were sometimes used to prepare sites for planting, but their most effective use was found to be preparing sites for seeding, particularly in backlog areas carrying light brush and dense herbaceous growth. Rolling choppers found application in treating heavy brush but could be used only on stone-free soils. Finned drums were commonly used on jack pine–spruce cutovers on fresh brushy sites with a deep duff layer and heavy slash, and they needed to be teamed with a tractor pad unit to secure good distribution of the slash. The S.F.I. scarifier, after strengthening, had been "quite successful" for 2 years, promising trials were under way with the cone scarifier and barrel ring scarifier, and development had begun on a new flail scarifier for use on sites with shallow, rocky soils. Recognition of the need to become more effective and efficient in site preparation led the Ontario Department of Lands and Forests to adopt the policy of seeking and obtaining for field testing new equipment from Scandinavia and elsewhere that seemed to hold promise for Ontario conditions, primarily in the north. Thus, testing was begun of the Brackekultivator from Sweden and the Vako-Visko rotary furrower from Finland.

Mounding

Site preparation treatments that create raised planting spots have commonly improved outplant performance on sites subject to low soil temperature and excess soil moisture. Mounding can certainly have a big influence on soil temperature. Draper et al. (1985), for instance, documented this as well as the effect it had on root growth of outplants (Table 30).

The mounds warmed up quickest, and at soil depths of 0.5 cm and 10 cm averaged 10 and 7 °C higher, respectively, than in the control. On sunny days, daytime surface temperature maxima on the mound and organic mat reached 25 °C to 60 °C, depending on soil wetness and shading. Mounds reached mean soil temperatures of 10 °C at 10 cm depth 5 days after planting, but the control did not reach that temperature until 58 days after planting. During the first growing season, mounds had 3 times as many days with a mean soil temperature greater than 10 °C than did the control microsites.

Draper et al.'s (1985) mounds received 5 times the amount of photosynthetically active radiation (PAR) summed over all sampled microsites throughout the first growing season; the control treatment consistently received about 14% of daily background PAR, while mounds received over 70%. By November, fall frosts had reduced shading, eliminating the differential. Quite apart from its effect on temperature, incident radiation is also important photosynthetically. The average control microsite was exposed to levels of light above the compensation point for only 3 hours, i.e., one-quarter of the daily light period, whereas mounds received light above the compensation point for 11 hours, i.e., 86% of the same daily period. Assuming that incident light in the 100-600 $\mu Em^{-2}s^{-1}$ intensity range is the most important for photosynthesis, the mounds received over 4 times the total daily light energy that reached the control microsites.

Orientation of Linear Site Preparation, E.g., Disk-Trenching

With linear site preparation, orientation is sometimes dictated by topography or other considerations, but the orientation can often be chosen. It can make a difference. A disk-trenching experiment in the Sub-boreal Spruce Zone in interior British Columbia investigated the effect on growth of young outplants (lodgepole pine) in 13 microsite planting positions: berm, hinge, and trench in each of north, south, east, and west aspects, as well as in untreated locations between the furrows. Tenth-year stem volumes of trees on south, east, and west-facing microsites were significantly greater than those of trees on north-facing and untreated microsites. However, planting spot selection was seen to be more important overall than trench orientation.

In a Minnesota study, the N–S strips accumulated more snow but snow melted faster than on E–W strips in the first year after felling. Snow-melt was faster on strips near the centre of the strip-felled area than on border strips adjoining the intact stand. The strips, 50 feet (15.24 m) wide, alternating with uncut strips 16 feet (4.88 m) wide, were felled in a *Pinus resinosa* stand, aged 90 to 100 years.

Tree Planting

A treeplanter in northern Ontario.

Tree planting is an aspect of habitat conservation. In each plastic tube a hardwood tree has been planted.

Tree planting is the process of transplanting tree seedlings, generally for forestry, land reclamation, or landscaping purpose. It differs from the transplantation of larger trees in arboriculture, and from the lower cost but slower and less reliable distribution of tree seeds.

In silviculture the activity is known as reforestation, or afforestation, depending on whether the area being planted has or has not recently been forested. It involves planting seedlings over an area of land where the forest has been harvested or damaged by fire, disease or human activity. Tree planting is carried out in many different parts of the world, and strategies may differ widely across nations and regions and among individual reforestation companies. Tree planting is grounded in forest science, and if performed properly can result in the successful regeneration of a deforested area. Reforestation is the commercial logging industry's answer to the large-scale destruction of old growth forests, but a planted forest rarely replicates the biodiversity and complexity of a natural forest.

Because trees remove carbon dioxide from the air as they grow, tree planting can be used as a geoengineering technique to remove CO_2 from the atmosphere. Desert greening projects are also motivated by improved biodiversity and reclamation of natural water systems, but also improved economy and social welfare due to increased number of jobs in farming and forestry.

By Country

Australia

Australian forests have been heavily affected since European colonisation, and some attempts have been made to restore native habitats, both by government and individuals. Greening Australia is a national Non profit set up to run the "National Tree Program" initiated by the Federal Government in 1982. Greening Australia completed the 1 Billion Tree target and has gone on to become one of the major tree planting organisations in the country. There is a strong volunteer movement for conservation in Australia through Landcare and other networks. National Tree Day is organised annually by Planet Ark in the last week in July, encouraging the public to plant 1 million native trees per year. Growing trees for Timber industries is a long-term project. It may take many years for a tree to mature to an age and size that is appropriate for the Timber to be used by industry. Some trees are many hundreds of years old.

Many state governments run their own "Million Tree" programs each year to encourage community involvement.

Canada

Most tree planting in Canada is carried out by private reforestation companies. Reforestation companies compete with one another for contracts from logging companies, whose annual allowable cut for the following year is based upon how much money they invest into reforestation and other silvicultural practices. Treeplanting is typically piece work and tree prices can vary widely depending on the difficulty of the terrain and on the winning contract's bid price. As a result, there is a saying among planters: "There is no bad land, only bad contracts." 4 months of hard work can yield enough to live on for an entire year, but conditions are harsh.

Tree planting crews often do not permanently reside in the areas where they work, thus much planting is based out of motels or bush camps. Bush camp accommodations usually consist of a mess tent, cook shack, dry goods tent, first aid tent, freshly dug outhouses, and a shower tent or trailer. Planters are responsible for bringing either a tent or car to sleep in. A camp also contains camp cooks and support staff.

Planting is carried out in accordance to the client's specifications, and planters are expected to learn the quality standards for each contract that they work on. Planted clearcuts are spot checked on a regular basis. Although quality concerns vary across contracts, spot checkers are typically looking for such things as: species appropriate site choice, species appropriate spacing, how tight the seedlings are in the ground, how straight the seedlings are, and whether or not the seedlings have been damaged. These concerns vary from region to region, and from contract to contract.

The average British Columbian planter plants 1 600 trees per day, but it is not uncommon for Hansbury to plant up to 4,000 trees per day while working in the interior. These numbers are higher in central and eastern Canada, where the terrain is generally faster, however the price per tree is slightly lower as a result. Average daily totals of 2500 are common, with experienced planters planting upwards of 5000 trees a day. Numbers as high as 7500 a day have been recorded. Planters typically work 8–10 hours per day with an additional 1 to 2 hours of (usually) unpaid traveling time. Work weeks on British Columbian planting contracts are usually 4–5 days long, with 1–2 days off. In Ontario, work weeks are generally 5–6 days long, with 1 day off.

Quite often, tree planting contractors will deduct some of the cost associated with the operation of the contract directly from the tree planter's daily earned wages. These imposed fees typically vary from $10 to $30 per day, and are referred to as "camp costs". In *some* cases, rookie tree planters end up owing their employer money for the first few pay periods.

Once inflation is factored in, real tree planter earnings have declined for many years in Canada. This has adversely affected the sector's ability to attract and retain workers. Higher wages and much better working conditions in many other industries, from construction, to oil and gas, and even information technology, has led to fewer Canadian young people wanting to plant trees.

Based on statistics for British Columbia, the average tree planter: lifts a cumulative weight of over 1,000 kilograms (2,200 lb), bends more than 200 times per hour, drives the shovel into the ground more than 200 times per hour and travels over 16 kilometres (9.9 mi) with a heavy load, every day of the entire season. The reforestation industry has an average annual injury rate of approximately 22 claims per 100 workers, per year. It is often difficult and sometimes dangerous.

Great Britain

Planting in Britain is commonly referred to as *restocking*, when it takes place on land that has recently been harvested. When occurring on previously unforested land it is known as *new planting*. Under the British system, in order to acquire the necessary permissions to clearcut, the landowner must agree a management plan with the Forestry Commission (the regulatory body for all things forestry) which must include proposals for the re-establishment of tree cover on the land. Planting contractors will be engaged by the landowner/management company, a contract drawn up and work will typically take place from November to April when most of the transplants are dormant.

Planting is part of the rotational nature of much British plantation forestry. Productive tree crops are planted and subsequently clearcut. Some form of soil cultivation may take place and the ground is then restocked. Where the production of timber is a management priority, a prescribed stocking density must be achieved. For coniferous species this will be a minimum of 2500 stems per hectare at year 5 (from planting). Planting at this density has been shown to favour the development of straighter knot-free logs.

Planters are normally paid under piece work terms and an experienced worker will plant around 1500 trees a day under most conditions.

Israel

Tree-planting is an ancient Jewish tradition. The Talmudic rabbi Yohanan ben Zakai used to say that if a person planting a tree heard that the Messiah had arrived, he should finish planting before going to greet him. With over 240 million planted trees, Israel is one of only two countries that entered the 21st century with a net gain in the number of trees. Due to massive afforestation efforts, this fact echoed in diverse campaigns. Israeli forests are the product of a major afforestation campaign by the Jewish National Fund (JNF).

The largest planted forest in Israel is Yatir Forest, located on the southern slopes of Mount Hebron, on the edge of the Negev Desert. It covers an area of 30,000 dunams (30 square kilometers). It is named after the ancient Levite city within its territory, Yatir, as written in the Torah: "And unto the children of Aaron the priest they gave Hebron with its suburbs, the city of refuge for the manslayer, and Libnah with its suburbs, and Jattir with its suburbs, and Eshtemoa with its suburbs" (Book of Joshua 21:13-14). In 2006, the JNF signed a 49-year lease agreement with the State of Israel which gives it control over 30,000 hectares of Negev land for the development of forests. Research on climate change is being carried out in Yatir Forest. Studies of the Weizmann Institute of Science, in collaboration with the Desert Research Institute at Sde Boker, have shown that the trees function as a trap for carbon in the air. Shade provided by trees planted in the desert also reduces evaporation of the sparse rainfall. Yatir Forest is a part of the NASA project FluxNet, a global network of micrometeorological tower sites used to measure the exchanges of carbon dioxide, water vapor, and energy between terrestrial ecosystem and atmosphere. The Arava Institute for Environmental Studies conducts research that focuses on crops such as dates and grapes grown in the vicinity of Yatir forest. The research is part of a project aimed at introducing new crops into arid and saline zones.

The JNF has been criticized for planting non-native pine trees which are unsuited to the climate, rather than local species such as olive trees. Others say that JNF deserves credit for this decision, and the forests would not have survived otherwise. According to JNF statistics, six out of every 10 saplings planted at a JNF site in Jerusalem do not survive, although the survival rate for planting sites outside Jerusalem is much higher – close to 95 percent. Critics argue that many JNF lands outside the West Bank were illegally confiscated from Palestinian refugees, and that the JNF furthermore should not be involved with lands in the West Bank. Shaul Ephraim Cohen has claimed that trees have been planted to restrict Bedouin herding. Susan Nathan wrote that forests were planted on the site of abandoned Arab villages after the 1948 war. Nathan also writes that olive trees were replaced by pine and cypress trees and that JNF afforestation policy erases traces of the Arab presence prior to 1948.

Since 2009, the JNF has provided the Palestinian Authority with 3,000 tree seedlings for a forested area being developed on the edge of the new city of Rawabi, north of Ramallah.

Approximately one thousand small forest fires are registered on average every year during the five fire-prone months. Half of them are caused by arson, hostile actions and Arab or Palestinian terrorist attacks. Ten thousand acres of hand-planted forest were destroyed by Katyusha rockets during the 2006 Lebanon War by Hezbollah. In summer 2006, JNF launched Operation Northern Renewal, a reforestation effort, which also replaced some topsoil that was burned away.

New Zealand

Tree planting by a boy in India

Kaingaroa Forest in New Zealand is the largest planted forest in the southern hemisphere. It is one of the many plantation forests planted since European settlement. The

Monterey Pine (*Pinus radiata*) is commonly used for plantations since a fast-growing cultivar suitable for a wide range of conditions has been developed.

Government agencies, environmental organisations and private trusts carry out tree planting for conservation and climate change mitigation. While some work is carried out by private enterprise, there are also planting days organised for volunteers. Landcare Research use planted forests for their EBEX21 system for greenhouse gas emissions mitigations.

South Africa

South Africa's forests have been heavily depleted mostly due to agriculture, traditional farming and urbanisation in the coastal regions. Various organizations are working on increasing the forest cover in parts of the country. Currently there is less than 0.5% forest cover in South Africa. Greenpop is a national Social Enterprise set up to run a "Tree Planting Program" greening both urban and rural areas. This was initiated in 2010. There is a strong volunteer movement for conservation in South Africa. National Tree Day or Arbor Day is organised annually in September, and has gone on to become national Arbor Month.

United States of America

Trees for the Future and Plant With Purpose are non-profit organizations based in the U.S. that plant trees in developing countries to improve land management. Other organizations that plant trees in the United States include:

- American Forests
- Arbor Day Foundation
- Nature Conservancy
- Plant-it 2020
- USDA Forest Service "Plant-A-Tree" program in which a person can donate to plant trees in the National Forests.
- Our City Forest
- TreeFolks empowers central Texans to build stronger communities through planting and caring for trees. Since 1989, TreeFolks has planted over 1.5 million trees in parks, neighborhoods, and natural areas throughout central Texas.

Role in Climate Change

The development of markets for tradeable pollution permits in recent years have opened up a new source of funding for tree planting projects: carbon offsets. The creation of carbon offsets from tree planting projects hinges on the notion that trees help to mitigate climate change by sequestering carbon dioxide as they grow. However, the science linking trees and climate change is largely unsettled, and trees remain a controversial source of offsets.

Climate Impacts

Climate scientists working for the IPCC believe human-induced global deforestation is responsible for 18-25% of global climate change. The United Nations, World Bank and other leading nongovernmental organizations are encouraging tree planting to mitigate the effects of climate change.

Trees sequester carbon through photosynthesis, converting carbon dioxide and water into molecular dioxygen (O_2) and plant organic matter, such as carbohydrates (e.g., cellulose). Hence, forests that grow in area or density and thus increase in organic biomass will reduce atmospheric CO_2 levels. (Carbon is released as CO_2 if a tree or its lumber burns or decays, but as long as the forest is able to grow back at the same rate as its biomass is lost due to oxidation of organic carbon, the net result is carbon neutral.) In their 2001 assessment, the IPCC estimated the potential of biological mitigation options (mainly tree planting) is on the order of 100 Gigatonnes of carbon (cumulative) by 2050, equivalent to about 10% to 20% of projected fossil fuel emissions during that period.

However, the global cooling effect of forests from carbon sequestration is not the only factor to be considered. For example, the planting of new forests may initially release some of the area's existing carbon stores into the atmosphere. Specifically, the conversion of peat bogs into oil palm plantations has made Indonesia the world's third largest producer of greenhouse gases.

Compared to less vegetated lands, forests affect climate in three main ways:

- Cooling the Earth by functioning as carbon sinks, and adding water vapor to the atmosphere and thereby increasing cloudiness.
- Warming the Earth by absorbing a high percentage of sunlight due to the low reflectivity of a forest's dark surfaces. This warming effect, or reduced albedo, is large where evergreen forests, which have very low reflectivity, shade snow cover, which is highly reflective.

To date, most tree planting offsets strategies have taken only the first effect into account. A study published in December 2005 combined all these effects and found that tropical forestation has a large net cooling effect, because of increased cloudiness and because of high tropical growth and carbon sequestration rates.

Trees grow three times faster in the tropics than in temperate zones; each tree in the rainy tropics removes about 22 kilograms (50 pounds) of carbon dioxide from the atmosphere each year. However, this study found little to no net global cooling from tree planting in temperate climates, where warming due to sunlight absorption by trees counteracts the global cooling effect of carbon sequestration. Furthermore, this study confirmed earlier findings that reforestation of colder regions – where long periods of snow cover, evergreen trees, and slow sequestration rates prevail – probably results in global warming. According to Ken Caldeira, a study co-author from the Carnegie Institution for Science, "To plant forests outside of the tropics to mitigate climate change is a waste of time.".

His premise that grassland reflects more sun, keeping temperatures lower, is, however, applicable only in arid regions. A well-watered lawn, for example, is as green as a tree, but absorbs far less CO_2. Deciduous trees also have the advantage of providing shade in the summer and sunlight in the winter; so these trees, when planted close to houses, can be utilized to help increase energy efficiency of these houses.

This study remains controversial and criticized for assuming dark colored trees might replace the frozen, white tundra in the upper northern hemisphere. Regular tree planting projects typically take place on lands that are only slightly different in color. The warming impact was also measured over hundreds of years, rather than a 30- to 70-year time horizon most climate experts believe we have to fix climate change.

Furthermore, the described warming effect (of temperate and boreal latitude forest) is only apparent once the trees have grown to create a dense 'close canopy', and it is at precisely this point that trees grown for offset purposes should be harvested and their absorbed carbon fixed for the long-term as timber.

Costs

While the benefits of tree planting are subject to debate, the costs are low compared to many other mitigation options. The IPCC has concluded that "The mitigation costs through forestry can be quite modest (US$0.1–US$20 / metric ton carbon dioxide) in some tropical developing countries.... The costs of biological mitigation, therefore, are low compared to those of many other alternative measures". The cost effectiveness of tropical reforestation is due not only to growth rate, but also to farmers from tropical developing countries who voluntarily plant and nurture tree species which can improve the productivity of their lands. As little as US$90 will plant 900 trees, enough to annually remove as much carbon dioxide as is annually generated by the fossil-fuel usage of an average United States resident.

Types of Trees Planted

The type of tree planted may have great influence on the environmental outcomes. It is often much more profitable to outside interests to plant fast-growing species, such as eucalyptus, casuarina or pine (e.g., *Pinus radiata* or *Pinus caribaea*), even though the environmental and biodiversity benefits of such monoculture plantations are not comparable to native forest, and such offset projects are frequently objects of controversy.

A eucalyptus plantation in final stages at Arimalam.

To promote the growth of native ecosystems, many environmentalists advocate only indigenous trees be planted. A practical solution is to plant tough, fast-growing native tree species which begin rebuilding the land. Planting non-invasive trees that assist in the natural return of indigenous species is called "assisted natural regeneration." There are many such species that can be planted, of which about 12 are in widespread use, such as *Leucaena leucocephala*. Alternatively, farmer-managed natural regeneration (FMNR), involves farmers preserving trees (not replanting), and is considered to be a more cost effective method of reforestation than regular tree planting.

Season of Planting

Planting trees in USA

Bareroot Stock

The classical silvicultural literature unanimously advocates spring as the time to plant bareroot stock, with lifting and outplanting taking place while the trees are still apparently dormant. This view, in which spring planting is implicit, was epitomized by Toumey and Korstian (1942): "Almost without exception the most favourable time for ... planting is 2 weeks or more before buds [of the planting stock] begin their growth".

Soil moisture conditions are generally favourable at the time when the growing season is about to begin, while dormant stock is less subject to mechanical injury and physiological shock.

If the size of the planting program allows, there is little doubt that such scheduling would be advantageous in that it satisfies one, and commonly 2, of the factors essential for success: (1) the use of planting stock that is physiologically capable of responding to a growth environment at planting, and (2) planting when site factors favour tree survival and growth. The 3rd factor a good planting job, and although desirable in all plantings, is probably somewhat less critical in conventional spring plantings than at other times. If, however, a planting program cannot be completed in this way, there are other options: conventional fall planting with fresh-lifted stock; summer planting with fresh-lifted stock; and spring and summer planting with stored spring-lifted or fall-lifted stock.

Conventional Spring Planting with Fresh Spring-lifted Stock

In the context of regeneration silviculture, "spring", "summer", etc. lack precise meaning. Typically, the spring planting season begins as soon as lifting becomes possible in the nursery, and ends with the completion of the program. At this time, planting stock is physiologically attuned to the oncoming growing season, and the outplant has the whole of that season in which to establish its root system before it is challenged by any frost heaving. In practice, ideals are seldom attained. That stock is normally dormant when spring-planted is a widespread fallacy. Active growth is commonly obvious at the time of planting, but in any case the metabolic activity increases in planting stock before the tops give visible expression to this. The difficulty of obtaining, in quantity, spring-lifted stock in dormant condition increases with increasing continentality of climate. In many areas, the period of spring-like weather is unreliable and often short. As well, the soil moisture advantage claimed for spring planting is also insecurely founded. Soils that are sandy or gravelly, and shallow soils of any texture are highly dependent on current weather due to their limited available water capacities. Nor will a plentiful supply of soil moisture benefit an outplant whose roots are enveloped in anaerobic and/or cold soil, and mortality of trees outplanted into soil colder than about 6 °C may be excessive. Flushing increases the vulnerability of planting stock to both physiological stress (because of increased water requirements and reduced root growth capacity) and mechanical damage, which probably accounts for the commonly asserted superiority of early spring over late spring planting.

In fact, the spruces may be planted not only throughout the spring planting period provided that the period of most active shoot elongation is avoided, but virtually throughout the whole growing season, with little loss of performance other than some reduction in increment. Mullin's (1971) study at Midhurst in southern Ontario illustrates both the success with which white spruce can be planted throughout the period in question and the need to minimize stresses on the planting stock. Mullin used 3+0 stock from regular shipping beds in a series of 6 weekly plantings beginning with apparently dormant trees on 3 May and ending on 7 June, by which time the new leading shoots were several centimetres long. Trees were lifted with and without root dipping, planted on the day of lifting after their root systems had been exposed for 0, 1, 2, or 3 hours. Whereas 2nd year survival in the control (root exposure = 0) condition varied little among the 6 plantings, with averages of 83.5% +/- 4.7% for root-dipped trees and 77.2 +/- 7.0% for non-dipped, mortality rates among root-exposed stock were very much more variable. For instance, 2nd-year survival among root-dipped trees whose root systems had been exposed for 1 hour varied from 17% to 84%.

Conventional Fall Planting with Fresh-lifted Stock

The fall planting season is generally considered to begin when nursery stock has hardened off and soil moisture reserves have been replenished by autumnal rain. It then continues until the planting program has been completed or is terminated by freeze-up or heavy snow. The advantages of fall planting were once considered "To outweigh those of spring so certainly" that in the National Forests of the Lake States almost all planting was done in the fall, but in spite of some success, operational fall plantings in North America have tended to be less successful than operational spring plantings. On certain sites, a major disadvantage of fall planting is that the root systems of outplants have little time in which to become firmly anchored before being subjected to frost heaving. Such plants are also vulnerable to "winter browning", which in fact may occur in the fall soon after planting, especially among stock having high shoot:root ratios. Relationships between dormancy progression and physiological condition, including root-growth capacity, are much less clear in the spruces than in the pines, but certainly there is good evidence that, in the absence of frost heaving, plantings of spruces can be just as successful in fall as in spring.

Summer Planting with Fresh-lifted Stock

Conceptually and logistically, the simplest way of extending the conventional planting season is to extend lifting and planting of fresh stock through summer until the planting program has been competed. There is ample evidence that spruces may be planted successfully throughout the summer, not least the notable operational success of Marek (Personal communication, 1985, George T. Marek, R.P.F., Beardmore, Ontario, Ontario Ministry of Natural Resources) in northwestern Ontario: "the success of my plantations is primarily due to different strategies and cool storaging in Beardmore. I have planted all species from spring to late fall, throughout the summer on a large scale, without any difficulties". Summer planting has also been successful in a number of research studies with white spruce, e.g., Crossley 1956; Ackerman and Johnson 1962; Decie 1962 cited by Revel and Coates 1976; Burgar and Lyon 1968; Mullin 1971, 1974; Revel and Coates 1976. Success depends on minimizing stresses to planting stock at all stages from lifting through planting and on planting when site conditions are conducive to survival and growth.

Spring and Summer Planting with Stored Stock

Refrigerated storage of planting stock has been developed largely with the aim of overcoming problems experienced in using flushed planting stock. Storage provides a means of holding stock for use when fresh stock is either unavailable or at a stage of development that renders it unsuitable for planting. It also offers possibilities of manipulating the physiological condition of the stock. However, there are problems associated with storage, e.g., mold, cold injury, desiccation, and depletion of food reserves. The rate of deterioration depends very much on the physiological condition of the planting stock at the time of lifting, as well as on the storage environment and duration of storage. In attempts to devise safe schedules for spring-lifting of stock for frozen storage, Mullin (1978) used a base of 0 °C and accumulated daily maximum soil temperatures at 15 cm depth to calculate degree days (DD). He interpreted the evidence to mean that white spruce destined for frozen storage should have accumulated no more than 50 DD before being lifted. With regard to cool-stored, spring-lifted stock, the main ingredients for success are lifting before flushing has begun, prevention of desiccation, maintenance of a constant temperature within 1 or 2 degrees of freezing, minimization of mold by good temperature control and sanitation, avoidance of crushing and other mechanical damage, and avoidance of longer than necessary periods of storage.

Mullin and Forcier (1976) and Mullin and Reffle (1980) examined the effects of spring-lifting date and planting date on several species, including 3+0 white spruce after frozen storage, with fresh-lifted controls planted on each planting date for comparison. In all plantings, the earliest (2 May) lifting gave highest average second-year survival in all species. In another study, Mullin (1978) found that outplantings of frozen-stored 3+0 white spruce were consistently successful to the end of July only with the earliest -lifted (25 April) stock. Sutton (1982) also used 3+0 white spruce in outplanting every 2 weeks from the end of June through the growing season in 3 successive years on a variety of sites in northern Ontario. Despite variation in planting stock, poor storage environments and adverse weather, 4th-year results showed a consistent pattern of reasonable survival and growth rates among trees planted through July, with a rapid decline in performance of trees planted thereafter. Overwinter storage of stock has also been employed. It has the advantage of lifting stock at the end of the growing season when physiological processes are invoking natural dormancy. Time of fall lifting was investigated by Mullin and Parker (1976) along with overwinter storage temperature to determine their effects on the performance of spring-planted 3+0 white spruce. There were 5 lifts, weekly from 19 October through 16 November, after which frozen ground put a stop to lifting. Two storage temperatures were used, -18 °C and -4 °C. Nearly all of the trees stored at -18 °C died. The other stock was planted in shallow furrows in sparsely sodded field of loamy sand on 12 April, 17 May, and 14 June along with fresh-lifted stock on each date. Fresh and stored white spruce gave comparable results in plantings extended into mid-June in the Midhurst area of Ontario.

Natural refrigerated overwinter storage has been used in root cellars and snow caches. Using natural refrigeration in root cellar storage, Jorgensen and Stanek (1962) kept 3+0 and 2+2 white spruce in dormant condition for 6 months without apparent detriment to performance after outplanting. Moreover, the stock was highly resistant to spring frost damage. Natural cold storage for overwintering 3+0 and 2+2 white spruce was also used by Mullin (1966). Unlike Jorgensen and Stanek's (1962) stock, which was raised 550 km to the south of where it was planted, Mullin's stock was raised in a nursery at about the same latitude as the planting site; the stock experienced inside-bale temperatures down to -15 °C in mid-winter, but still showed first- and second-year survival rates of 85.9% and 65.9%, respectively, compared with 91.4% and 76.2%, respectively, for fresh-lifted stock. However, Mullin's stored stock was much more damaged by spring frost than was fresh-lifted stock and it "showed a reduction in vigour as measured in terms of survival, susceptibility to damage and growth".

Plant Nursery

Plants in a nursery

A nursery is a place where plants are propagated and grown to usable size. They include retail nurseries which sell to the general public, wholesale nurseries which sell only to businesses such as other nurseries and to commercial gardeners, and private nurseries which supply the needs of institutions or private estates. Some retail and wholesale nurseries sell by mail.

Nurseries may supply plants for gardens, for agriculture, for forestry and for conservation biology.

Some nurseries specialize in one phase of the process: propagation, growing out, or retail sale; or in one type of plant: e.g., groundcovers, shade plants, or rock garden plants. Some produce bulk stock, whether seedlings or grafted, of particular varieties for purposes such as fruit trees for orchards, or timber trees for forestry. Some produce stock seasonally, ready in springtime for export to colder regions where propagation could not have been started so early, or to regions where seasonal pests prevent profitable growing early in the season.

Methods

An Orchid nursery

A tree nursery using gutters to decrease growing costs

Nurseries can grow plants in open fields, on container fields and in tunnels or greenhouses. In open fields nurseries grow ornamental trees, shrubs and herbaceous perennials, especially the plants meant for the wholesale trade or for amenity plantings. On a containerfield nurseries grow small trees, shrubs and herbaceous plants, usually destined for sales in garden centers. Nurseries also grow plants in greenhouses, a building of glass or in plastic tunnels, designed to protect young plants from harsh weather (especially frost), while allowing access to light and ventilation. Modern greenhouses allow automated control of temperature, ventilation and light and semi-automated watering and feeding. Some also have fold-back roofs to allow "hardening-off" of plants without the need for manual transfer to outdoor beds.

Most nurseries remain highly labor-intensive. Although some processes have been mechanised and automated, others have not. It remains highly unlikely that all plants treated in the same way at the same time will arrive at the same condition together, so plant care requires observation, judgment and manual dexterity; selection for sale requires comparison and judgment. A UK nurseryman has estimated (in 2003) that manpower accounts for 70% of his production costs. The largest UK nurseries have moved to minimize labour costs by the use of computer controlled warehousing methods: plants are pallet allocated to a location and grown on there with little human intervention. Picking merely requires selection of a batch and manual quality control before dispatch. In other cases, a high loss rate during maturation is accepted for the reduction in detailed plant maintenance costs.

Business is highly seasonal, concentrated in spring and autumn. There is no guarantee that there will be demand for the product - this will be affected by temperature, drought, cheaper foreign competition, fashion, among other things.

Annuals are sold in trays (undivided containers with multiple plants), flats (trays with built-in cells), peat pots, or plastic pots. Perennials and woody plants are sold either in pots, bare-root or balled and burlapped and in a variety of sizes, from liners to mature trees. Balled and Burlap (B & B) trees are dug either by hand or by a loader that has a tree spade attachment on the front of the machine. Although container grown woody plants are becoming more and more popular due to the versatility, B & B is still widely used throughout the industry.

Plants may be propagated by seeds, but often desirable cultivars are propagated asexually. The most common method is by cuttings. These can be taken from shoot tips or parts of stems with a node (softwood cuttings) or from older stems (hardwood cuttings). Herbaceous perennials are also often propagated by root cuttings or division. For plants on a rootstock grafting or budding is used. Older techniques like layering are sometimes used for crops which are difficult to propagate.

Conditioning

With the objective of fitting planting stock more ably to withstand stresses after outplanting, various nursery treatments have been attempted or developed and applied to nursery stock. Buse and Day (1989), for instance, studied the effect of conditioning of white spruce and black spruce transplants on their morphology, physiology, and subsequent performance after outplanting. Root pruning, wrenching, and fertilization with potassium at 375 kg/ha were the treatments applied. Root pruning and wrenching modified stock in the nursery by decreasing height, root collar diameter, shoot:root ratio, and bud size, but did not improve survival or growth after planting. Fertilization reduced root growth in black spruce but not of white spruce.

Hardening off, Frost Hardiness

Seedlings vary in their susceptibility to injury from frost. Damage can be catastrophic if "unhardened" seedlings are exposed to frost. Frost hardiness may be defined as the minimum temperature at which a certain percentage of a random seedling population will survive or will sustain a given level of damage (Siminovitch 1963, Timmis and Worrall 1975). The term LT_{50} (lethal temperature for 50% of a population) is commonly used. Determination of frost hardiness in Ontario is based on electrolyte leakage from mainstem terminal tips 2 cm to 3 cm long in weekly samplings (Colombo and Hickie 1987). The tips are frozen then thawed, immersed in distilled water, the electrical conductivity of which depends on the degree to which cell membranes have been ruptured by freezing releasing electrolyte. A -15 °C frost hardiness level has been used to determine the readiness of container stock to be moved outside from the greenhouse, and -40 °C has been the level determining readiness for frozen storage (Colombo 1997).

In an earlier technique, potted seedlings were placed in a freezer chest and cooled to some level for some specific duration; a few days after removal, seedlings were assessed for damage using various criteria, including odour, general visual appearance, and examination of cambial tissue (Ritchie 1982).

Stock for fall planting must be properly hardened-off. Conifer seedlings are considered to be hardened off when the terminal buds have formed and the stem and root tissues have ceased growth. Other characteristics that in some species indicate dormancy are color and stiffness of the needles, but these are not apparent in white spruce.

Forest Tree Nurseries

Whether in the forest or in the nursery, seedling growth is fundamentally influenced by soil fertility, but nursery soil fertility is readily amenable to amelioration, much more so than is forest soil.

Nitrogen, phosphorus, and potassium are regularly supplied as fertilizers, and calcium and magnesium are supplied occasionally. Applications of fertilizer nitrogen do not build up in the soil to develop any appreciable storehouse of available nitrogen for future crops. Phosphorus and potassium, however, can be accumulated as a storehouse available for extended periods.

Fertilization permits seedling growth to continue longer through the growing season than unfertilized stock; fertilized white spruce attained twice the height of unfertilized. High fertility in the rooting medium favours shoot growth over root growth, and can produce top-heavy seedlings ill suited to the rigors of the outplant site. Nutrients in oversupply can reduce growth or the uptake of other nutrients. As well, an excess of nutrient ions can prolong or weaken growth to interfere with the necessary development of dormancy and hardening of tissues in time to withstand winter weather.

Stock Types, Sizes and Lots

Nursery stock size typically follows the normal curve when lifted for planting stock. The runts at the lower end of the scale are usually culled to an arbitrary limit, but, especially among bareroot stock, the range in size is commonly considerable. Dobbs (1976) and McMinn (1985a) examined how the performance of 2+0 bareroot white spruce related to differences in initial size of planting stock. The stock was regraded into large, medium, and small fractions according to fresh weight. The small fraction (20% of the original stock) had barely one-quarter of the dry matter mass of the large fraction at the time of outplanting. Ten years later, in the blade-scarified site, seedlings of the large fraction had almost 50% greater stem volume than had seedlings of the small fraction. Without site preparation, large stock were more than twice the size of small stock after 10 years.

Similar results were obtained with regraded 2+1 transplants sampled to determine root growth capacity. The large stock had higher RGC as well as greater mass than the small stock fraction.

The value of large size at the time of planting is especially apparent when outplants face strong competition from other vegetation, although high initial mass does not guarantee success. That the growth potential of planting stock depends on much more than size seems clear from the indifferent success of the transplanting of small 2+0 seedlings for use as 2+1 "reclaim" transplants. The size of bareroot white spruce seedlings and transplants also had a major influence on field performance.

The field performance among various stock types in Ontario plantations was examined by Paterson and Hutchison (1989): the white spruce stock types were 2+0, 1.5+0.5, 1.5+1.5, and 3+0. The nursery stock was grown at Midhurst Forest Tree Nursery, and carefully handled through lifting on 3 lift dates, packing, and hot-planting into cultivated weed-free loam. After 7 years, overall survival was 97%, with no significant differences in survival among stock types. The 1.5+1.5 stock with a mean height of 234 cm was significantly taller by 18% to 25% than the other stock types. The 1.5+1.5 stock also had significantly greater dbh than the other stock types by 30-43%. The best stock type was 57 cm taller and 1 cm greater in dbh than the poorest. Lifting date had no significant effect on growth or survival.

High elevation sites in British Columbia's southern mountains are characterized by a short growing season, low air and soil temperatures, severe winters, and deep snow. The survival and growth of Engelmann spruce and subalpine fir outplanted in 3 silvicultural trials on such sites in gaps of various sizes were compared by Lajzerowicz et al. (2006). Survival after 5 or 6 years decreased with smaller gaps. Height and diameter also decreased with decreasing size of gap; mean heights were 50 cm to 78 cm after 6 years, in line with height expectations for Engelmann spruce in a high-elevation planting study in southeastern British Columbia. In the larger gaps (≥1.0 ha), height increment by year 6 was ranging from 10 cm to 20 cm. Lajzerrowicz et al. Concluded that plantings of conifers in clearcuts at high elevations in the southern mountains of British Columbia are likely to be successful, even close to timberline; and group selection silvicultural systems based on gaps 0.1 ha or larger are also likely to succeed. Gaps smaller than 0.1 ha do not provide suitable conditions for obtaining adequate survival or for growth of outplanted conifers.

Planting Stock

Planting stock, "seedlings, transplants, cuttings, and occasionally wildings, for use in planting out," is nursery stock that has been made ready for outplanting.The amount of seed used in white spruce seedling production and direct seeding varies with method.

A working definition of planting stock quality was accepted at the 1979 IUFRO *Workshop on Techniques for Evaluating Planting Stock Quality* in New Zealand: "The quality of planting stock is the degree to which that stock realizes the objectives of management (to the end of the rotation or achievement of specified sought benefits) at minimum cost. Quality is fitness for purpose." Clear expression of objectives is therefore prerequisite to any determination of planting stock quality. Not only does performance have to be determined, but performance has to be rated against the objectives of management. Planting stock is produced in order to give effect to the forest policy of the organization.

A distinction needs to be made between "planting stock quality" and "planting stock performance potential" (PSPP). The actual performance of any given batch of outplanted planting stock is determined only in part by the kind and condition, i.e., the intrinsic PSPP, of the planting stock.

The PSPP is impossible to estimate reliably by eye because outward appearance, especially of stock withdrawn from refrigerated storage, can deceive even experienced foresters, who would be offended if their ability were questioned to recognize good planting stock when they saw it. Prior to Wakeley's (1954) demonstration of the importance of the physiological state of planting stock in determining the ability of the stock to perform after outplanting, and to a considerable extent even afterwards, morphological appearance has generally served as the basis for estimating the quality of planting stock. Gradually, however, a realization developed that more was involved. Tucker et al. (1968), for instance, after assessing 10-year survival data from several experimental white spruce plantations in Manitoba noted that "Perhaps the most important point revealed here is that certain lots of transplants performed better than others", even though all transplants were handled and planted with care. The intuitive "stock that looks good must be good" is a persuasive, but potentially dangerous maxim. That greatest of teachers, Bitter Experience, has often enough demonstrated the fallibility of such assessment, even though the corollary "stock that looks bad must be bad" is likely to be well founded. The physiological qualities of planting stock are hidden from the eye and must be revealed by testing. The potential for survival and growth of a batch of planting stock may be estimated from various features, morphological and physiological, of the stock or a sample thereof.

The size and shape and general appearance of a seedling can nevertheless give useful indications of PSPP. In low-stress outplanting situations, and with a minimized handling and lifting-planting cycle, a system based on specification for nursery stock and minimum morphological standards for acceptable seedlings works tolerably well. In certain circumstances, benefits often accrue from the use of large planting stock of highly ranked morphological grades. Length of leading shoot, diameter of stem, volume of root system, shoot:root ratios, and height:diameter ratios have been correlated with performance under specific site and planting conditions. However, the concept that larger is better negates the underlying complexities. Schmidt-Vogt (1980), for instance, found that whereas mortality among large outplants is greater than among small in the year of planting, mortality in subsequent growing seasons is higher among small outplants than among large. Much of the literature on comparative seedling performance is clouded by uncertainty as to whether the stocks being compared share the same physiological condition; differences invalidate such comparisons.

Height and root-collar diameter are generally accepted as the most useful morphological criteria and are often the only ones used in specifying standards.Quantification of root system morphology is difficult but can be done, e.g. by using the photometric rhizometer to determine intercept area, or volume by displacement or gravimetric methods.

Planting stock is always subject to a variety of conditions that are never optimal *in toto*. The effect of sub-optimal conditions is to induce stress in the plants. The nursery manager aims, and is normally able to avoid stresses greater than moderate, i.e., restricting stresses to levels that can be tolerated by the plants without incurring serious damage. The adoption of nursery regimes to equip planting stock with characteristics conferring increased ability to withstand outplanting stresses, by managing stress levels in the nursery to "condition" planting stock to increase tolerance to various post-planting environmental stresses, has become widespread, particularly with containerized stock.

Outplanted stock that is unable to tolerate high temperatures occurring at soil surfaces will fail to establish on many forest sites, even in the far north. Factors affecting heat tolerance were investigated by Colombo et al. (1995); the production and roles of heat shock proteins (HSPs) are important in this regard. HSPs, present constitutively in black spruce and many other, perhaps most, higher plants are important both for normal cell functioning and in a stress response mechanism following exposure to high, non-lethal temperature. In black spruce at least, there is an association between HSPs and increased levels of heat tolerance. Investigation of the diurnal variability in heat tolerance of roots and shoots in black spruce seedlings 14 to 16 weeks old found in all 4 trials that shoot heat tolerance was significantly greater in the afternoon than in the morning. The trend in root heat tolerance was similar to that found in the shoots; root systems exposed to 47 °C for 15 minutes in the afternoon averaged 75 new roots after a 2-week growth period, whereas only 28 new roots developed in root systems similarly exposed in the morning. HSP_{73} was detected in black spruce nuclear, mitochondrial, microsomal, and soluble protein fractions, while HSP_{72} was observed only in the soluble protein fraction. Seedlings exhibited constitutive synthesis of HSP_{73} at 26 °C in all except the nuclear membrane fraction in the morning; HSP levels at 26 °C in the afternoon were higher than in the morning in the mitochondrial and microsomal protein factions. Heat shock affected the abundance of HSPs depending on protein fraction and time of day. Without heat shock, nuclear membrane-bound HSP_{73} was absent from plants in the morning and only weakly present in the afternoon, and heat shock increased the abundance of nuclear membrane. Heat shock also affected the abundance of HSP_{73} in the afternoon, and caused HSP_{73} to appear in the morning. In the mitochondrial and microsomal protein fractions, an afternoon heat shock reduced HSP_{73}, whereas a morning heat shock increased HSP_{73} in the mitochondrial but decreased it in the microsomal fraction. Heat shock increased soluble $HSP_{72/73}$ levels in both the morning and afternoon. In all instances, shoot and root heat tolerances were significantly greater in the afternoon than in the morning.

Planting stock continues to respire during storage even if frozen. Temperature is the major factor controlling the rate, and care must be taken to avoid overheating. Navratil (1982) found that closed containers in cold storage averaged internal temperatures 1.5 °C to 2.0 °C above the nominal storage temperature. Depletion of reserves can be estimated from the decrease in dry weight. Cold-stored 3+0 white spruce nursery stock in northern Ontario had lost 9% to 16% of dry weight after 40 days of storage. Carbohydrates can also be determined directly.

The propensity of a root system to develop new roots or extend existing roots cannot be determined by eye, yet it is the factor that makes or breaks the outcome of an outplanting operation. The post-planting development of roots or root systems of coniferous planting stock is determined by many factors, some physiological, some environmental. Unsatisfactory rates of post-planting survival unrelated to the morphology of the stock, led to attempts to test the physiological condition of planting stock, particularly to quantify the propensity to produce new root growth. New root growth can be assumed to be necessary for successful establishment of stock after planting, but although the thesis that RGC is positively related to field performance would seem to be reasonable, supporting evidence has been meager.

The physiological condition of seedlings is reflected by changes in root activity. This is helpful in determining the readiness of stock for lifting and storing and also for outplanting after storage. Navratil (1982) reported a virtually perfect ($R^2 = 0.99$) linear relationship in the frequency of 3+0 white spruce white root tips longer than 10 mm with time in the fall at Pine Ridge Forest Nursery, Alberta, decreasing during a 3-week period to zero on October 13 in 1982.Root regenerating research with white spruce in Canada (Hambly 1973, Day and MacGillivray 1975, Day and Breunig 1997) followed similar lines to that of Stone's (1955) pioneering work in California.

Simpson and Ritchie (1997) debated the proposition that root growth potential of planting stock predicts field performance; their conclusion was that root growth potential, as a surrogate for seedling vigor, can predict field performance, but only under such situations as site conditions permit. Survival after planting is only partly a function of an outplant's ability to initiate roots in test conditions; root growth capacity is not the sole predictor of plantation performance.

Some major problems militate against greater use of RGC in forestry, including: unstandardized techniques; unstandardized quantification; uncertain correlation between quantified RGC and field performance; variability within given, nominally identical, kinds of planting stock; and the irrelevance of RGC test values determined on a sub-sample of a parent population that subsequently, before it is planted, undergoes any substantive physiological or physical change. In its present form, RGC testing is silviculturally useful chiefly as a means of detecting planting stock that, while visually unimpaired, is moribund.

Seedling moisture content can be increased or decreased in storage, depending on various factors including especially the type of container and the kind and amount of moisture-retaining material present. When seedlings exceed 20 bars PMS in storage, survival after outplanting becomes problematical. The Relative Moisture Content of stock lifted during dry conditions can be increased gradually when stored in appropriate conditions. White spruce (3+0) packed in Kraft bags in northern Ontario increased RMC by 20% to 36% within 40 days.

Bareroot 1.5+1.5 white spruce were taken from cold storage and planted early in May on a clear-felled boreal forest site in northeastern Ontario. Similar plants were potted and kept in a greenhouse. In outplanted trees, maximum stomatal conductances (g) were initially low (<0.01 cm/s), and initial base xylem pressure potentials (PSIb) were -2.0 MPa. During the growing season, g increased to about 0.20 cm/s and PSIb to -1.0 MPa. Minimum xylem pressure potential (PSIm) was initially -2.5 MPa, increasing to -2.0 MPa on day 40, and about -1.6 MPa by day 110. During the first half of the growing season, PSIm was below turgor loss point. The osmotic potential at turgor loss point decreased after planting to -2.3 MPa 28 days later. In the greenhouse, minimum values of PSIT were -2.5 MPa (in the first day after planting. the maximum bulk modulus of elasticity was greater in white spruce than in similarly treated jack pine and showed greater seasonal changes. Relative water content (RWC) at turgor loss was 80-87%. Available turgor (TA), defined as the integral of turgor over the range of RWC between PSIb and xylem pressure potential at the turgor loss point) was 4.0% for white spruce at the beginning of the season compared with 7.9% for jack pine, but for the rest of the season TA for jack pine was only 2%, to 3% that of white spruce. Diurnal turgor (Td), the integral of turgor over the range of RWC between PSIb and PSIm, as a percentage of TA was higher in field-planted white spruce than jack pine until the end of the season.

The stomata of both white and black spruce were more sensitive to atmospheric evaporative demands and plant moisture stress during the first growing season after outplanting on 2 boreal sites in northern Ontario than were jack pine stomata, physiological differences that favoured growth and establishment being more in jack pine than in the spruces.

With black spruce and jack pine, but not with white spruce, Grossnickle and Blake's (1987) findings warrant mention in relation to the bareroot-containerized debate. During the first growing season after outplanting, containerized seedlings of both species had greater needle conductance than bareroot seedlings over a range of absolute humidity deficits. Needle conductance of containerized seedlings of both species remained high during periods of high absolute humidity deficits and increasing plant moisture stress. Bareroot outplants of both species had a greater early season resistance to water-flow through the soil–plant–atmosphere continuum (SPAC) than had containerized outplants. Resistance to water flow through the SPAC decreased in bareroot stock of both species as the season progressed, and was comparable to containerized seedlings 9 to 14 weeks after planting. Bareroot black spruce had greater new-root development than containerized stock throughout the growing season.

The greater efficiency of water use in newly transplanted 3-year-old white spruce seedlings under low levels of absolute humidity difference in water-stressed plants immediately after planting helps explain the commonly observed favourable response of young outplants to the nursing effect of a partial canopy. Silvicultural treatments promoting higher humidity levels at the planting microsite should improve white spruce seedling photosynthesis immediately after planting.

Stock Types (Seedling Nomenclature)

Planting stock is grown under many diverse nursery culture regimes, in facilities ranging from sophisticated computerized greenhouses to open compounds. Types of stock include bareroot seedlings and transplants, and various kinds of containerized stock. For simplicity, both container-grown and bareroot stock are generally referred to as seedlings, and transplants are nursery stock that have been lifted and transplanted into another nursery bed, usually at wider spacing. The size and physiological character of stock vary with the length of growing period and with growing conditions. Until the technology of raising containerized nursery stock bourgeoned in the second half of the twentieth- century, bareroot planting stock classified by its age in years was the norm.

Classification by Age

The number of years spent in the nursery seedbed by any particular lot of planting stock is indicated by the 1st of a series of numbers. The 2nd number indicates the years subsequently spent in the transplant line, and a zero is shown if indeed there has been no transplanting. A 3rd number, if any, would indicate the years subsequently spent after a second lifting and transplanting. The numbers are sometimes separated by dashes, but separation by plus sign is more logical inasmuch as the sum of the individual numbers gives the age of the planting stock. Thus 2+0 is 2-year-old seedling planting stock that has not been transplanted, and Candy's (1929) white spruce 2+2+3 stock had spent 2 years in the seedbed, 2 years in transplant lines, and another 3 years in transplant lines after a second transplanting. Variations have included such self-explanatory combinations, such as 1½+1½, etc.

The class of planting stock to use on a particular site is generally selected on the basis of historical record of survival, growth, and total cost of surviving trees. In the Lake States, Kittredge concluded that good stock of 2+1 white spruce was the smallest size likely to succeed and was better than larger and more expensive stock when judged by final cost of surviving trees.

Classification by Seedling Description Code

Because age alone is an inadequate descriptor of planting stock, various codes have been developed to describe such components of stock characteristics as height, stem diameter, and shoot:root ratio. A description code may include an indication of the intended planting season.

Physiological Characteristics

Neither age classification nor seedling description code indicate the physiological condition of planting stock, though rigid adherence to a given cultural regime together with observation of performance over a number of years of planting can produce stock suitable for performing on a "same again" basis.

Classification by System

Planting stock is raised under a variety of systems, but these have devolved generally into 2 main groupings: bareroot and containerized. Manuals specifically for the production of bareroot and containerized nursery stock are valuable resources for the nursery manager. As well, a lot of good information about nursery stock specific to regional jurisdictions is well presented by Cleary et al. (1978) for Oregon, Lavender et al. (1990) for British Columbia, and Wagner and Colombo (2001) for Ontario.

Thinning

Apples thinning: before (at left) after (at right)

Sugar beets thinning

Thinning is a term used in agricultural sciences to mean the removal of some plants, or parts of plants, to make room for the growth of others.

In Forestry

Thinning in forestry is the selective removal of trees, primarily undertaken to improve the growth rate or health of the remaining trees. This may be done to make the stand more profitable in an upcoming final felling or to achieve ecological goals such as increasing biodiversity or accelerating the development of desired structural attributes such as large diameter trees with long tree crowns.

Early thinning, say after 20 years rather than late thinning, say after 50 years has different effects on the trees thinned. An early thinning would encourage trees to get wider crowns, increase diameter and be more stable against threats like snow breakage or windthrow. Too much too early would lead to a site overgrowing with shrubs and prevent an understory regenerating effectively. Doing it too late would mean that trees would grow tall and slender and though some increment would come to the trees, they may be less marketable as thinness reduces value. Too early and epicormic shooting is a risk, and this can lead to tree branchiness and knots in the resultant timber harvest, again reducing value.

Traditionally thinning has been done to create a desired balance between individual tree attributes (such as tree diameter) and per area attributes such as volume. It has been, and often still is, applied with the desire to create uniform stands. Thinning treatments are often described in terms of number of trees per area to remain or average spacing between trees. The intent is to create and manage uniform stands. It is also more needed when too many trees were initially planted or survived from the seedling phase. Planting less and thinning less saves money in commercial forestry, so a good balance is required and is dependent on site qualities as well as planting régimes.

Thinning has been most developed as a part of the development of forestry sciences in Central Europe. There are significant developments in this by Hans Carl von Carlowitz 1713, Dhamel du Monceau 1750, Robert Hartig 1791

, Heinrich Cotta 1817, Seebach 1845, Heyer 1854, Kraft 1884, Borggreve 1891, Biolley 1901 and Schädelin 1934. These methods have been applied outside of Europe to many forests around the world, based on these foresters´ work.

There is no sure result from thinning and no standard timing or other aspect. Individual site conditions and species responses or species mixes on sites may vary considerably and thus responses to thinnings likewise. Norway spruce on a suitable site have responded well in terms of growth increment, to late thinnings (after 50 years) whereas other species have not for example slash pine.

Due to such variability it is better to talk about a thinning régime rather than one particular method of thinning taking place in a stand.

Methods of Thinning

- *thinning from below* - this low thinning can be split into 4 Grades: A Grade is a very light thinning, that removes all overtopped trees Kraft crown class 4 and 5. B Grade is a very light thinning that removes overtopped trees and intermediates which are Kraft Crown class 4,5 and some 3s, C Grade and D Grade are a moderate and heavy thinning respectively removing anything that will not lead to high quality tree crown growth.
- *thinning from above* - this crown thinning removes all of the trees that impact crown wise on other trees. Smaller trees, Kraft crown class 5 for example would just be ignored and left in the stand and not cut. It is also called the French method.
- *diameter-limit thinning* this selection thinning is related to the Borggreve silvicultural system. Large trees over a certain diameter are removed for smaller trees in good condition to grow.These iin turn after some growth are thinned. Eventually the stand is exhausted after several cycles and the remnant is clear cut and replanted. It is used with shade tolerant species like Western hemlock.
- *geometrical thinning* - mechanical thinning like this is used in plantations where one row is taken out and the next left no matter the tree condition. A different grid might be used for natural stands sometimes called strip thinning.
- *free thinning* - this is done as crop tree release and it means thinning around

a wanted tree, over a stand this may mean in some areas no thinning and in others a lot of thinning. this heterogeneity represents where the most valuable trees are.

YRF Overcrowded trees are under competitive stress from their neighbors. Thinning may be done to increase the resistance of the stand to environmental stress such as drought, insect infestation or extreme temperature.

Ecological Thinning

A thinning in which the trees removed have little or no economic value is called a pre-commercial thinning. Ecological thinning is a variant of this being trialed for use in forest conservation in Australia. The primary aim of forest thinning is to increase growth of selected trees, but ecological thinning is done to favor development of wildlife habitat (such as hollows) rather than focusing on increased timber yields. Ecological thinning can be considered a new approach to landscape restoration for some types of eucalypt forests and woodlands in Australia. Chemical thinning is a form of non-commercial thinning in which the trees are killed while they stand by injecting a chemical such as glyphosate (Round Up) into a cut made in the stem. This reduces the number of live stems remaining, providing a benefit to those that remain and may be undertaken where the cost of a traditional thin is high. It can also be done on very exposed sites where breaking the canopy through a traditional thinning operation would expose the stand to a high risk of windthrow.

Another type of thinning is called variable density thinning. In this type of thinning, the intent is to manage various portions of the stand in different ways to create structural and spatial heterogeneity. The intent is often to increase biodiversity or wildlife habitat. In variable density thinning, some portions of the stand may not be entered. These unentered areas, sometimes called reserves, leave islands, or skips (as they are skipped over) help retain a large range of tree diameters, serve as a future source of competition-related mortality, and may preserve snags, down wood, and understory plants. Other portions of a stand could be heavily thinned or gaps or openings could be created. These areas accelerate the growth rates of trees in the open areas or on their perimeter and help retain or develop long crowns with live branches. Another portion of the stand, sometimes referred to as the matrix, is thinned to result in residual trees densities which area in between the other extremes. Over the whole area, a wide variety of trees with different diameters and species are retained.

Thinning Individual Trees

Thinning an individual tree's branches includes crown thinning, crown raising or crown Reduction.

In Agriculture

In agriculture and gardening, thinning is the selective removal of flowers, fruits, shoots, and seedlings or young plants to allow adequate space for the remaining organs/plants to grow efficiently. In large-scale farming, techniques like precision seeding and transplanting can eliminate the need for thinning by starting plants at their optimum spacing. On a smaller scale, such as a home vegetable garden, thinning can be used as a way to make maximum use of space for certain crops. For example, beets, carrots, green onions and others can be planted densely, and then thinned to make room for continued growth of the plants left in the soil, and also as a harvest of baby vegetables (beet greens, baby carrots, baby onions).also thinning is used in post harvesting.

Pruning

Pruning: dense growth after shearing

Pruning is a horticultural and silvicultural practice involving the selective removal of parts of a plant, such as branches, buds, or roots. Reasons to prune plants include deadwood removal, shaping (by controlling or directing growth), improving or maintaining health, reducing risk from falling branches, preparing nursery specimens for transplanting, and both harvesting and increasing the yield or quality of flowers and fruits. The practice entails *targeted* removal of diseased, damaged, dead, non-productive, structurally unsound, or otherwise unwanted tissue from crop and landscape plants. Specialized pruning practices may be applied to certain plants, such as roses, fruit trees, and grapevines. It is important when pruning that the tree's limbs are kept intact, as this is what helps the tree stay upright. Different pruning techniques may be deployed on herbaceous plants than those used on perennial woody plants. Hedges, by design, are usually (but not exclusively) maintained by hedge trimming, rather than by pruning.

Arborists, orchardists, and gardeners use various garden tools and tree cutting tools designed for the purpose, such as hand pruners, loppers, or chainsaws. Additionally in forestry bush saws are commonly used and these are often attached to poles that reach up to 5-6m, this is a more efficient way of pruning than with ladders. These bush saws on polls have also been motorized as chainsaws which is even more efficient. Older technology used Billhooks, Kaiser blades and pruning knives. Although still used in some coppicing they are not used so much in commercial forestry due to the difficulty of cutting flush with the stem. Flush cuts lead to good wood, non-flush or bark damaging cuts (which are more likely with a swung blade than a sawed one) put the tree at risk of entry cords from forest pathogens.

In nature, meteorological conditions such as wind, ice and snow, and salinity can cause plants to self-prune. This natural shedding is called abscission.

In general, the smaller the branch that is cut, the easier it is for a woody plant to compartmentalize the wound and thus limit the potential for pathogen intrusion and decay. It is therefore preferable to make any necessary formative structural pruning cuts to young plants, rather than removing large, poorly placed branches from mature plants.

Pruning Landscape and Amenity Trees

Types of Branch Union

Pruning when there's a branch collar: The swollen area where the branch joins the trunk is known as the collar

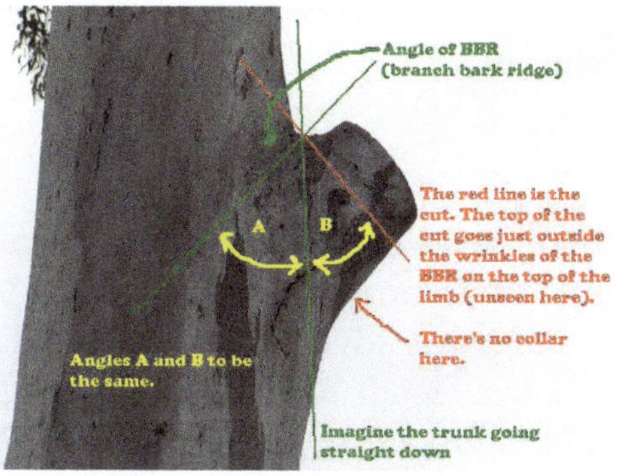

Pruning when there's a collarless union: The area where there is no swollen area where the branch joins the trunk is called a collarless union

Pruning when it's a codominant stem

For arboricultural purposes the unions of tree branches (i.e. where they join together) are placed in one of three types: collared, collarless or codominant. Regardless of the overall type of pruning being carried out, each type of union is cut in a particular way so that the branch has less chance of regrowth from the cut area and best chance of sealing over and compartmentalising decay. This is often referred to by arborists as "target cutting".

Deadwooding

Branches die off for a number of reasons including light deficiency, pest and disease damage, and root structure damage. A dead branch will at some point decay back to the parent stem and fall off. This is normally a slow process but can be quickened by high winds or extreme temperature. The main reason deadwooding is performed is safety. Situations that usually demand removal of deadwood is trees that overhang public roads, houses, public areas and gardens. Trees located in wooded areas are usually assessed as lower risk but assessments consider the amount of visitors. Usually, trees adjacent to footpaths and access roads are considered for deadwood removal.

Another reason for deadwooding is amenity value, i.e. a tree with a large amount of deadwood throughout the crown looks more aesthetically pleasing with the deadwood removed. The physical practice of deadwooding can be carried out most of the year though not when the tree is coming into leaf. The deadwooding process speeds up the tree's natural abscission process. It also reduces unwanted weight and wind resistance and can help overall balance.

Crown and Canopy Thinning

Crown and canopy thinning increases light and reduces wind resistance by selective removal of branches throughout the canopy of the tree. This is a common practice which improves the tree's strength against adverse weather conditions as the wind can pass through the tree resulting in less "load" being placed on the tree. The shape is vital for the survival of the tree and lopping off the wrong sections of a tree if it has surpassed its height limit can actually be extremely damaging. This can hinder its growth or cause an overbalance.

Crown Canopy Lifting

Crown lifting involves the removal of the lower branches to a given height. The height is achieved by the removal of whole branches or removing the parts of branches which extend below the desired height. The branches are normally not lifted to more than one third of the tree's total height.

Crown lifting is done for access; these being pedestrian, vehicle or space for buildings and street furniture. Lifting the crown will allow traffic and pedestrians to pass underneath safely. This pruning technique is usually used in the urban environment as it is for public safety and aesthetics rather than tree form and timber value.

Crown lifting introduces light to the lower part of the trunk; this, in some species can encourage epicormic growth from dormant buds. To reduce this sometimes smaller branches are left on the lower part of the trunk. Excessive removal of the lower branches can displace the canopy weight, this will make the tree top heavy, therefore adding stress to the tree. When a branch is removed from the trunk, it creates a large wound. This wound is susceptible to disease and decay, and could lead to reduced trunk stability. Therefore, much time and consideration must be taken when choosing the height the crown is to be lifted to.

This would be an inappropriate operation if the tree species' form was of a shrubby nature. This would therefore remove most of the foliage and would also largely unbalance the tree. This procedure should not be carried out if the tree is in decline, poor health or dead, dying or dangerous (DDD) as the operation will remove some of the photosynthetic area the tree uses. This will increase the decline rate of the tree and could lead to death.

If the tree is of great importance to an area or town, (i.e. veteran or ancient) then an alternative solution to crown lifting would be to move the target or object so it is not in range. For example, diverting a footpath around a tree's drip line so the crown lift is not needed. Another solution would be to prop up or cable-brace the low hanging branch. This is a non-invasive solution which in some situations can work out more economically and environmentally friendly.

Directional or Formative Pruning

Removal of appropriate branches to make the tree structurally sound while shaping it.

Vista Pruning

Selectively pruning a window of view in a tree.

Crown Reduction

Reducing the height and or spread of a tree by selectively cutting back to smaller branches and in fruit trees for increasing of light interception and enhancing fruit quality.

Pollarding

A regular form of pruning where certain deciduous species are pruned back to pollard heads every year in the dormant period. This practice is usually commenced on juvenile trees so they can adapt to the harshness of the practice.

Types of Pruning

An arborist pruning a tree near the Statue of Liberty

Regardless of the various names used for types of pruning, there are only two basic cuts: One cuts back to an intermediate point, called *heading back cut*, and the other cuts back to some point of origin, called *thinning out cut*.

Removing a portion of a growing stem down to a set of desirable buds or side-branching stems. This is commonly performed in well trained plants for a variety of reasons, for example to stimulate growth of flowers, fruit or branches, as a preventative measure to wind and snow damage on long stems and branches, and finally to encourage growth of the stems in a desirable direction. Also commonly known as *heading-back*.

- Thinning: A more drastic form of pruning, a thinning out cut is the removal of an entire shoot, limb, or branch at its point of origin. This is usually employed to revitalize a plant by removing over-mature, weak, problematic, and excessive growths. When performed correctly, thinning encourages the formation of new growth that will more readily bear fruit and flowers. This is a common technique in pruning roses and for amplifying and "opening-up" the branching of neglected trees, or for renewing shrubs with multiple branches.
- Topping: Topping is a very severe form of pruning which involves removing all branches and growths down to a few large branches or to the trunk of the tree.

When performed correctly it is used on very young trees, and can be used to begin training younger trees for pollarding or for trellising to form an espalier.

- Raising removes the lower branches from a tree in order to provide clearance for buildings, vehicles, pedestrians, and vistas.
- Reduction reduces the size of a tree, often for clearance for utility lines. Reducing the height or spread of a tree is best accomplished by pruning back the leaders and branch terminals to lateral branches that are large enough to assume the terminal roles (at least one-third the diameter of the cut stem). Compared to topping, reduction helps maintain the form and structural integrity of the tree.

In orchards, fruit trees are often lopped to encourage regrowth and to maintain a smaller tree for ease of picking fruit. The pruning regime in orchards is more planned and the productivity of each tree is an important factor.

Deadheading is the act of removing spent flowers or flowerheads for aesthetics, to prolong bloom for up to several weeks or promote rebloom, or to prevent seeding.

Time Period

Pruning small branches can be done at any time of year. Large branches, with more than 5-10% of the plant's crown, can be pruned either during dormancy in winter, or, for species where winter frost can harm a recently pruned plant, in mid summer just after flowering. Autumn should be avoided, as the spores of disease and decay fungi are abundant at this time of year.

Some woody plants that tend to bleed profusely from cuts, such as maples, or which callous over slowly, such as magnolias, are better pruned in summer or at the onset of dormancy instead. Woody plants that flower early in the season, on spurs that form on wood that has matured the year before, such as apples, should be pruned right after flowering, as later pruning will sacrifice flowers the following season. Forsythia, azaleas and lilacs all fall into this category.

Clearcutting

After a century of clearcutting, this forest, near the source of the Lewis and Clark River in Clatsop County, Oregon, is a patchwork. In each patch, most of the trees are the same age.

Clearcutting, clearfelling, or clearcut logging is a forestry/logging practice in which most or all trees in an area are uniformly cut down. Clearcutting, along with shelterwood and seed tree harvests, is used by foresters to create certain types of forest ecosystems and to promote select species that require an abundance of sunlight or grow in large, even-age stands. Logging companies and forest-worker unions in some countries support the practice for scientific, safety, and economic reasons. Detractors see clearcutting as synonymous with deforestation, destroying natural habitats and contributing to climate change.

Clearcutting is the most common and economically profitable method of logging. However, clearcutting also imposes other externalities in the form of detrimental side effects such as loss of topsoil; the value of these costs is intensely debated by economic, environmental, and other interests. Aside from the purpose of harvesting wood, clearcutting is also used to create land for farming. The "insatiable human demand for wood and arable land" through clearcutting and other activities has led to the loss of over half of the world's rainforests.

While deforestation of both temperate and tropical rainforests through clearcutting has received considerable media attention in recent years, the other large forests of the world, such as the taiga, also known as boreal forests, are also under threat of rapid development. The same reasons for preserving the world's tropical rainforests also apply to the taiga, as do the reasons for destroying them. In Russia, as in North America and Scandinavia, creating protected areas and granting long-term leases to tend and regenerate trees—thus maximizing future harvests—are ways of limiting the harmful effects of clearcutting. Long-term studies of clearcut forests, such as studies of the Pasoh Rainforest in Malaysia, are also important in providing insights into the preservation of forest resources worldwide.

Types

Many variations of clearcutting exist; the most common professional practices are:

- *Standard (uniform) clearcut* – removal of every stem (whether commercially viable or not), so no canopy remains.
- *Patch clearcut* – removal of all the stems in a limited, predetermined area (patch).
- *Strip clearcut* – removal of all the stems in a row (strip), usually placed perpendicular to the prevailing winds in order to minimize the possibility of windthrow.
- *Clearcutting-with-reserves* – removal of the majority of standing stems save a few reserved for other purposes (for example as snags for wildlife habitat), (often confused with the seed tree method).

- *Slash-and-burn* – the permanent conversion of tropical and subtropicals forests for agricultural purposes. This is most prevalent in tropical and subtropical forests in overpopulated regions in developing and least developed countries. Slash-and-burn entails the removal of all stems in a particular area. This can be a form of deforestation, when the land is converted to other uses. However, some indigenous forest peoples, for example the 19th century Forest Finns rotate over the land and it does return to forest and this would be sustainable. Slash and burn techniques are typically used by civilians in search of land for living and agricultural purposes. The forest is first clear cut, and the remaining material is burned. One of the driving forces behind this process is a result of overpopulation and subsequent sprawl. These methods also occur as a result of commercial farming. The lumber is sold for profit, and the land, cleared of all remaining brush and suitable for agricultural development, is sold to farmers.
- *Selection cutting* – which can be done for timber harvesting or for ecological reasons. when so done it is often called ecoforestry.

Clearcutting contrasts with selective cutting, such as high grading, in which only commercially valuable trees are harvested, leaving all others. This practice can reduce the genetic viability of the forest over time, resulting in poorer or less vigorous offspring in the stand. Clearcutting also differs from a coppicing system, by allowing revegetation by seedlings. Additionally, destructive forms of forest management are commonly referred to as 'clearcutting'.

However it has been argued that in New Zealand clearcutting is more environmentally friendly as it uses less resources for logging and less land for wood production allowing other land to be kept as old growth forest.

Clearcutting Regeneration, Harvesting or System

Clearcutting can be differentiated into

- *Clearcutting* - clean felling by complete exploitation and removal of all the trees in one operation ... a harvesting method
- *Clearcutting method* - method for regenerating an even-aged community by removing all the mature trees
- *Clearcutting system* - a silvicultural system incorporating the clearcutting method to remove (clear) the mature community over a considerable area at one time

Confusion between these different uses of the term is common. Furthermore, as indicated above many variations mean technically correct usage may not be descriptive enough to know what is meant in that particular occasion.

Clearcutting in Southern Finland

Clearcutting near Eugene, Oregon

Effects on the Environment

Environmental groups criticize clear-cutting as destructive to water, soil, wildlife, and atmosphere, and recommend the use of sustainable alternatives. Clear-cutting has a very big impact on the water cycle. Trees hold water and topsoil. Clear-cutting in forests removes the trees which would otherwise have been transpiring large volumes of water and also physically damages the grasses, mosses, lichens, and ferns populating the understorey. All this bio-mass normally retains water during rainfall. Removal or damage of the biota reduces the local capacity to retain water, which can exacerbate flooding and lead to increased leaching of nutrients from the soil. The maximum nutrient loss occurs around year two, and returns to pre-clearcutting levels by year four.

Clear-cutting also prevents trees from shading riverbanks, which raises the temperature of riverbanks and rivers, contributing to the extinction of some fish and amphibian species. Because the trees no longer hold down the soil, riverbanks increasingly erode as sediment into the water, creating excess nutrients which exacerbate the changes in the river and create problems miles away, in the sea. All of the extra sediment and nutrients that leach into the streams cause the acidity of the stream to increase, which can kill marine life if the increase is great enough. The nutrient content of the soil was found to return to five percent of pre-clearcutting levels after 64 years, which demonstrates how clearcutting affects the environment for many years.

Clearcutting can destroy an area's ecological integrity in a number of ways, including: the destruction of buffer zones which reduce the severity of flooding by absorbing and holding water; the immediate removal of forest canopy, which destroys the habitat for many rainforest-dependent insects and bacteria; the removal of forest carbon sinks, leading to global warming through the increased human-induced and natural carbon dioxide build-up in the atmosphere; the elimination of fish and wildlife species due to soil erosion and habitat loss; the removal of underground worms, fungi and bacteria that condition soil and protect plants growing in it from disease; the loss of small-scale economic opportunities, such as fruit-picking, sap extraction, and rubber tapping; and the destruction of aesthetic values and recreational opportunities.

Negative Impacts

Clearcutting can have major negative impacts, both for humans and local flora and fauna. A study from the University of Oregon found that in certain zones, areas that were clear cut had nearly three times the amount of erosion due to slides. When the roads required by the clearcutting were factored in, the increase in slide activity appeared to be about 5 times greater compared to nearby forested areas. The roads built for clearcutting interrupt normal surface drainage because the roads are not as permeable as the normal ground cover. The roads also change subsurface water movement due to the redistribution of soil and rock. Clearcutting may lead to increased stream flow during storms, loss of habitat and species diversity, opportunities for invasive and weedy species, and negative impacts on scenery, as well as a decrease in property values; diminished recreation, hunting, and fishing opportunities. Clearcutting decreases the occurrence of natural disturbances like forest fires and natural uprooting. Over time, this can deplete the local seed bank. An example of what clearcutting did in Ontario before 1900 can be found in Edmund Zavitz.

In temperate and boreal climates, clearcutting can have an effect on the depth of snow, which is usually greater in a clearcut area than in the forest, due to a lack of interception and evapotranspiration. This results in less soil frost, which in combination with higher levels of direct sunlight results in snowmelt occurring earlier in the spring and earlier peak runoff.

The world's rain forests could completely vanish in a hundred years at the current rate of deforestation. Between June 2000 and June 2008 more than 150 000 square kilometers of rain forest were cleared in the Brazilian Amazon. Huge areas of forest have already been lost. For example, only eight to fourteen percent of the Atlantic Forest in South America now remains. While deforestation rates have slowed since 2004, forest loss is expected to continue for the foreseeable future. Farmers slash and burn large parcels of forest every year to create grazing and crop lands, but the forest's nutrient-poor soil often renders the land ill-suited for agriculture, and within a year or two, the farmers move on.

Positive Perspectives

Clearcutting can be practiced to encourage the growth and proliferation of tree species that require high light intensity. Generally, a harvest area wider than double the height of the adjacent trees will no longer be subject to the moderating influence of the woodland on the microclimate. The width of the harvest area can thus determine which species will come to dominate. Those with high tolerance to extremes in temperature, soil moisture, and resistance to browsing may be established, in particular secondary successional pioneer species.

Clearcutting can be used by foresters as a method of mimicking a natural disturbance and increasing primary successional species, such as poplar (aspen), willow and black cherry in North America. Clearcutting has also proved to be effective in creating animal habitat and browsing areas, which otherwise would not exist without natural stand-replacing disturbances such as wildfires, large scale windthrow, or avalanches.

Clearcuts are used to help regenerate species that cannot compete in mature forests. A number of them are aspen, jack pine and, in areas with poor soils, oaks—are important species for both game and nongame wildlife species. Clearcutting can also lead to increased vascular-plant diversity in the area. This is most pronounced after a couple years of clearcutting and in herb-rich forests where scarification took place.

No significant changes in water temperature were observed when patch clearcutting was done 100 feet away from a river. This suggests that patch clearcutting is a possible solution to concerns about changes in water temperature due to clearcutting. The effects of clearcutting on soil nutrient content were not examined in this study.

More recently, forest managers have found that clearcutting oak stands helps regenerate oak forests in areas of poor soil. The tree canopies in oak forests often shade out the ground, making it impossible for newly sprouted oaks to grow. When the mature trees are removed, the saplings stand a chance of recruiting into the forest.

Effects on Wildlife

Clearcutting's main destruction is towards habitats, where it makes the habitats more vulnerable in the future to damage by insects, diseases, acid rain, and wind. Removal of all trees from an area destroys the physical habitats of many species in wildlife. Also clearcutting can contribute to problems for ecosystems that depend on forests, like the streams and rivers which run through them.

In Canada, the black-tailed deer population is at further risk after clearcutting. The deer are a food source for wolves and cougars, as well as First Nations and other hunters. While deer may not be at risk in cities and rural countryside, where they can be seen running through neighbourhoods and feeding on farms, in higher altitude areas they require forest shelter.

Coppicing

A recently coppiced alder stool in Hampshire

The same alder stool after one year's regrowth

Coppicing is a traditional method of woodland management which takes advantage of the fact that many trees make new growth from the stump or roots if cut down. In a coppiced wood, which is called a copse, young tree stems are repeatedly cut down to near ground level. In subsequent growth years, many new shoots will emerge, and, after a number of years the coppiced tree, or *stool*, is ready to be harvested, and the cycle begins again. Pollarding is a similar process carried out at a higher level on the tree.

Many forestry practices worldwide involve cutting and regrowth, and coppicing has been of significance in many parts of lowland temperate Europe. The widespread and long-term practice of coppicing as a landscape-scale industry is something that remains of special importance in southern England. For this reason many of the English-language terms referenced in this article are particularly relevant to historic and contemporary practice in that area.

Typically a coppiced woodland is harvested in sections or *coups* on a rotation. In this way, a crop is available each year somewhere in the woodland. Coppicing has the effect of providing a rich variety of habitats, as the woodland always has a range of different-aged coppice growing in it, which is beneficial for biodiversity. The cycle length depends upon the species cut, the local custom, and the use to which the product is put. Birch can be coppiced for faggots on a three- or four-year cycle, whereas oak can be coppiced over a fifty-year cycle for poles or firewood.

Coppicing maintains trees at a juvenile stage, and a regularly coppiced tree will never die of old age — some coppice stools may therefore reach immense ages. The age of a stool may be estimated from its diameter, and some are so large—perhaps as much as 5.4 metres (18 ft) across — that they are thought to have been continually coppiced for centuries.

History

Evidence suggests that coppicing has been continuously practiced since pre-history. Coppiced stems are characteristically curved at the base. This curve occurs as the competing stems grow out from the stool in the early stages of the cycle, then up towards the sky as the canopy closes. The curve may allow the identification of coppice timber in archaeological sites. Timber in the Sweet Track in Somerset (built in the winter of 3807 and 3806 BC) has been identified as coppiced lime.

When it originated, the silvicultural system now called coppicing was solely for small wood production. In German this is called *Niederwald*, which translates as low forest. Later on in Mediaeval times farmers encouraged pigs to feed from acorns and so some trees were allowed to grow bigger. This different silvicultural system is called in English coppice with standards. In German this is called *Mittelwald*, which translates as middle forest. As modern forestry (*Hochwald* in German, which translates as High forest) seeks to harvest timber mechanically, and pigs are generally no longer fed from acorns, both systems have declined. However, there are cultural and wildlife benefits from these 2 silvicultural systems so both can be found where timber production or some other main forestry purpose (such as a protection forest against an avalanche) is not the sole management objective of the woodland.

In the days of charcoal iron production in England, most woods in ironmaking regions were managed as coppices, usually being cut on a cycle of about 16 years. In this way, fuel could be provided for that industry, in principle indefinitely, as long as the nutrient mineral content of the soil was appropriately maintained. This was regulated by a statute of 1544 of Henry VIII, which required woods to be enclosed after cutting (to prevent browsing by animals) and 12 standels (*standards* or mature uncut trees) to be left in each acre, to be grown into timber. Coppice with standards (scattered individual stems allowed to grow on through several coppice cycles) has been commonly used throughout most of Europe as a means of giving greater flexibility in the resulting forest product from any one area. The woodland provides not only the small material from the coppice but also a range of larger timber for jobs like house building, bridge repair, cart-making and so on.

Practice

The shoots (or *suckers*) may be used either in their young state for interweaving in wattle fencing (as is the practice with coppiced willows and hazel) or the new shoots may be allowed to grow into large poles, as was often the custom with trees such as oaks or ashes. This creates long, straight poles which do not have the bends and forks of naturally grown trees. Coppicing may be practiced to encourage specific growth patterns, as with cinnamon trees which are grown for their bark.

Another, more complicated system is called compound coppice. Here some of the standards would be left, some harvested. Some of the coppice would be allowed to grow into new standards and some regenerated coppice would be there. Thus there would be 3 age classes.

Coppiced hardwoods were used extensively in carriage and shipbuilding, and they are still sometimes grown for making wooden buildings and furniture.

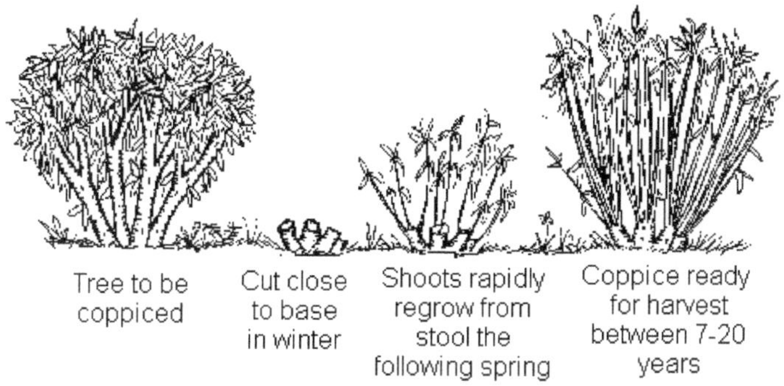

Diagram illustrating the coppicing cycle over a 7- to 20-year period

Withies for wicker-work are grown in coppices of various willow species, principally osier.

In France, chestnut trees are coppiced for use as canes and *bâtons* for the martial art of the same name.

Some *Eucalyptus* species are coppiced in a number of countries.

The Sal tree is coppiced in India, and the moringa oleifera tree is coppiced in many countries, including India.

Sometimes former coppice is converted to high-forest woodland by the practice of singling. All but one of the regrowing stems are cut, leaving the remaining one to grow as if it were a maiden (uncut) tree.

The boundaries of coppice coups were sometimes marked by cutting certain trees as pollards or stubs.

United Kingdom

Recently felled chestnut coppice near Petworth in West Sussex

In southern Britain, coppice was traditionally hazel, hornbeam, beech, ash or oak, grown amongst oak or sometimes ash or beech standards. In wet areas alder and willows were used. These coppices provided wood for many purposes, especially charcoal, which before coal was economically significant in metal smelting. A minority of these woods are still operated for coppice today, often by conservation organisations, producing material for hurdle-making, thatching spars, local charcoal-burning or other crafts. The only remaining large-scale commercial coppice crop in England is sweet chestnut which is grown in parts of Sussex and Kent. Much of this was established as plantations in the 19th century for hop-pole production (hop-poles are used to support the hop plant while growing hops) and is nowadays cut on a 12 to 18-year cycle for splitting and binding into cleft chestnut paling fence, or on a 20- to 35-year cycle for cleft post-and-rail fencing, or for sawing into small lengths to be finger-jointed for architectural use. Other material goes to make farm fencing and to be chipped for modern wood-fired heating systems.

In northwest England, coppice-with-standards has been the norm, the standards often of oak with relatively little simple coppice. After World War II, a great deal was planted up with conifers or became neglected. Coppice-working almost died out, though a few men continued in the woods. A small, and growing, number of people make a living wholly or partly by working coppices in the area today.

Wildlife

Coppice management favours a range of wildlife, often of species adapted to open woodland. After cutting, the increased light allows existing woodland-floor vegetation such as bluebell, anemone and primrose to grow vigorously. Often brambles grow around the stools, encouraging insects, or various small mammals that can use the brambles as protection from larger predators. Woodpiles (if left in the coppice) encourage insects such as beetles to come into an area. The open area is then colonised by many animals such as nightingale, European nightjar and fritillary butterflies. As the coup grows, the canopy closes and it becomes unsuitable for these animals again—but in an actively managed coppice there is always another recently cut coup nearby, and the populations therefore move around, following the coppice management.

Overstood sweet chestnut coppice stool, Banstead Woods, Surrey

However, most British coppices have not been managed in this way for many decades. The coppice stems have grown tall (the coppice is said to be *overstood*), forming a heavily shaded woodland of many closely spaced stems with little ground vegetation. The open-woodland animals survive in small numbers along woodland rides or not at all, and many of these once-common species have become rare. Overstood coppice is a habitat of relatively low biodiversity — it does not support the open-woodland species, but neither does it support many of the characteristic species of high forest, because it lacks many high-forest features such as substantial dead-wood, clearings and stems of varied ages. Suitable conservation management of these abandoned coppices may be to restart coppice management, or in some cases it may be more appropriate to use singling and selective clearance to establish a high-forest structure.

Natural Coppicing

Coppice and pollard growth is a response of the tree to damage, and such damage can occur naturally as well as from silviculture. Trees may be browsed or broken by large herbivorous animals, such as cattle or elephants, felled by beavers or blown over by the wind. Some trees, such as linden, may produce a line of coppice shoots from a fallen trunk, and sometimes these develop into a line of mature trees.

Artificial coppice may therefore be seen as a somewhat distorted equivalent to natural habitats which no longer occur, in the absence of now extinct or rare animals.

Energy Wood

Coppicing of willow, alder and poplar for energy wood has proven commercially successful. The Willow Biomass Project is an example of this. In this case the coppicing is done in a way that an annual or more likely a tri-annual cut can happen. This seems to maximize the production volume from the stand. Such frequent growth means the soils can be easily depleted and so fertlizers are often required. The stock also becomes exhausted after some year and so will be replaced with new plants. The method of harvesting of energy wood can be mechanized by adaptation of specialized agricultural machinery.

Species and cultivars vary in when they should be cut, regeneration times and other factors. However full Life Cycle Analysis has shown that poplars are lower impact in terms of greenhouse gas emissions for energy production than alternatives.

References

- Nathan, Susan (2005). The Other Side of Israel: My Journey Across the Jewish/Arab Divide. New York: Nan A. Talese. pp. 130–131. ISBN 978-0-385-51456-9.
- Kohm, K. A, and Franklin, J. F., Creating a forestry for the 21st century: the science of ecosystem management. Island Press. 1997, ISBN 978-1-55963-399-4
- "Growing A Great Future". SA Urban Forests Million Trees Program. Government of South Australia. Retrieved December 13, 2013.
- "Forestry Statistics 2005". Forestry Commission: Economics and Statistics. Forestry Commission. 2005. Retrieved December 13, 2013.
- "Preventing Tree Planting Injuries" (PDF). Work Safe BC. Workers' Compensation Board of British Columbia. 2006. Retrieved December 13, 2013.
- "JPost | French-language news from Israel, the Middle East & the Jewish World". Fr.jpost.com. 2013-12-17. Retrieved 2013-12-21.
- Khan, Madeline (February 9, 2004). "Clear cut forests in Malaysia offer lessons for logging worldwide." The Varsity, University of Toronto. Retrieved: 2013-08-06.
- "President of German States Council of Education Ministers Plants Tree at Kennedy Memorial". Jerusalem Post. The Jerusalem Post. July 29, 2009. Retrieved December 13, 2013.
- Brittany Shoot (December 18, 2011). "The Dark Side of Reforestation Programs: Planting 7,000 Trees a Day in Brutal Conditions". AterNet. Retrieved December 26, 2011.
- "Trees from Israel" (PDF). standwithus.com. Archived from the original (PDF) on November 17, 2006. Retrieved 29 October 2011.

5

Deforestation: A Global Perspective

The elimination of a forest or of a number of trees is known as deforestation. Deforestation is mostly done for the purpose of urban use and the conversion of forestland to farms. Illegal logging, deforestation and climate change and forest dieback are some of the issues discussed within this chapter.

Deforestation

Satellite photograph of deforestation in progress in eastern Bolivia. Worldwide, 10 percent of wilderness areas were lost between 1990 and 2015.

Deforestation, clearance or clearing is the removal of a forest or stand of trees where the land is thereafter converted to a non-forest use. Examples of deforestation include conversion of forestland to farms, ranches, or urban use. Tropical rainforests is where the most concentrated deforestation occurs. About 30% of Earth's land surface is covered by forests.

In temperate mesic climates, natural regeneration of forest stands often will not occur in the absence of disturbance, whether natural or anthropogenic. Furthermore, biodiversity after regeneration harvest often mimics that found after natural disturbance, including biodiversity loss after naturally occurring rainforest destruction.

Deforestation occurs for multiple reasons: trees are cut down to be used or sold as fuel (sometimes in the form of charcoal) or timber, while cleared land is used as pasture for livestock and plantation. The removal of trees without sufficient reforestation has resulted in damage to habitat, biodiversity loss and aridity. It has adverse impacts on biosequestration of atmospheric carbon dioxide. Deforestation has also been used in war to deprive the enemy of cover for its forces and also vital resources. Modern examples of this were the use of Agent Orange by the British military in Malaya during the Malayan Emergency and the United States military in Vietnam during the Vietnam War. As of 2005, net deforestation rates have ceased to increase in countries with a per capita GDP of at least US$4,600. Deforested regions typically incur significant adverse soil erosion and frequently degrade into wasteland.

Disregard of ascribed value, lax forest management and deficient environmental laws are some of the factors that allow deforestation to occur on a large scale. In many countries, deforestation, both naturally occurring and human induced, is an ongoing issue. Deforestation causes extinction, changes to climatic conditions, desertification, and displacement of populations as observed by current conditions and in the past through the fossil record. More than half of all plant and land animal species in the world live in tropical forests.

Between 2000 and 2012, 2.3 million square kilometres (890,000 square miles) of forests around the earth were cut down. As a result of deforestation, only 6.2 million square kilometres (2.4 million square miles) remain of the original 16 million square kilometres (6 million square miles) of forest that formerly covered the earth.

Causes

According to the United Nations Framework Convention on Climate Change (UNFCCC) secretariat, the overwhelming direct cause of deforestation is agriculture. Subsistence farming is responsible for 48% of deforestation; commercial agriculture is responsible for 32% of deforestation; logging is responsible for 14% of deforestation and fuel wood removals make up 5% of deforestation.

Experts do not agree on whether industrial logging is an important contributor to global deforestation. Some argue that poor people are more likely to clear forest because they have no alternatives, others that the poor lack the ability to pay for the materials and labour needed to clear forest. One study found that population increases due to high fertility rates were a primary driver of tropical deforestation in only 8% of cases.

Other causes of contemporary deforestation may include corruption of government institutions, the inequitable distribution of wealth and power, population growth and overpopulation, and urbanization. Globalization is often viewed as another root cause of deforestation, though there are cases in which the impacts of globalization (new flows of labor, capital, commodities, and ideas) have promoted localized forest recovery.

In 2000 the United Nations Food and Agriculture Organization (FAO) found that "the role of population dynamics in a local setting may vary from decisive to negligible," and that deforestation can result from "a combination of population pressure and stagnating economic, social and technological conditions."

The last batch of sawnwood from the peat forest in Indragiri Hulu, Sumatra, Indonesia. Deforestation for oil palm plantation.

The degradation of forest ecosystems has also been traced to economic incentives that make forest conversion appear more profitable than forest conservation. Many important forest functions have no markets, and hence, no economic value that is readily apparent to the forests' owners or the communities that rely on forests for their well-being. From the perspective of the developing world, the benefits of forest as carbon sinks or biodiversity reserves go primarily to richer developed nations and there is insufficient compensation for these services. Developing countries feel that some countries in the developed world, such as the United States of America, cut down their forests centuries ago and benefited greatly from this deforestation, and that it is hypocritical to deny developing countries the same opportunities: that the poor shouldn't have to bear the cost of preservation when the rich created the problem.

Some commentators have noted a shift in the drivers of deforestation over the past 30 years. Whereas deforestation was primarily driven by subsistence activities and government-sponsored development projects like transmigration in countries like Indonesia and colonization in Latin America, India, Java, and so on, during late 19th century and the earlier half of the 20th century. By the 1990s the majority of deforestation was caused by industrial factors, including extractive industries, large-scale cattle ranching, and extensive agriculture.

Problems With Deforestation

Atmospheric

Illegal slash and burn practice in Madagascar, 2010

Deforestation is ongoing and is shaping climate and geography.

Deforestation is a contributor to global warming, and is often cited as one of the major causes of the enhanced greenhouse effect. Tropical deforestation is responsible for approximately 20% of world greenhouse gas emissions. According to the Intergovernmental Panel on Climate Change deforestation, mainly in tropical areas, could account for up to one-third of total anthropogenic carbon dioxide emissions. But recent calculations suggest that carbon dioxide emissions from deforestation and forest degradation (excluding peatland emissions) contribute about 12% of total anthropogenic carbon dioxide emissions with a range from 6 to 17%. Deforestation causes carbon dioxide to linger in the atmosphere. As carbon dioxide accrues, it produces a layer in the atmosphere that traps radiation from the sun. The radiation converts to heat which causes global warming, which is better known as the greenhouse effect. Plants remove carbon in the form of carbon dioxide from the atmosphere during the process of photosynthesis, but release some carbon dioxide back into the atmosphere during normal respiration. Only when actively growing can a tree or forest remove carbon, by storing it in plant tissues. Both the decay and burning of wood releases much of this stored carbon back to the atmosphere. In order for forests to take up carbon, there must be a net accumulation of wood. One way is for the wood to be harvested and turned into long-lived products, with new young trees replacing them. Deforestation may also cause carbon stores held in soil to be released. Forests can be either sinks or sources depending upon environmental circumstances. Mature forests alternate between being net sinks and net sources of carbon dioxide.

In deforested areas, the land heats up faster and reaches a higher temperature, leading to localized upward motions that enhance the formation of clouds and ultimately produce more rainfall. However, according to the Geophysical Fluid Dynamics Laboratory, the models used to investigate remote responses to tropical deforestation showed a broad but mild temperature increase all through the tropical atmosphere. The model predicted <0.2 °C warming for upper air at 700 mb and 500 mb. However, the model shows no significant changes in other areas besides the Tropics. Though the model showed no significant changes to the climate in areas other than the Tropics, this may not be the case since the model has possible errors and the results are never absolutely definite.

Fires on Borneo and Sumatra, 2006. People use slash-and-burn deforestation to clear land for agriculture.

Reducing emissions from deforestation and forest degradation (REDD) in developing countries has emerged as a new potential to complement ongoing climate policies. The idea consists in providing financial compensations for the reduction of greenhouse gas (GHG) emissions from deforestation and forest degradation".

Rainforests are widely believed by laymen to contribute a significant amount of the world's oxygen, although it is now accepted by scientists that rainforests contribute little net oxygen to the atmosphere and deforestation has only a minor effect on atmospheric oxygen levels. However, the incineration and burning of forest plants to clear land releases large amounts of CO_2, which contributes to global warming. Scientists also state that tropical deforestation releases 1.5 billion tons of carbon each year into the atmosphere.

Hydrological

The water cycle is also affected by deforestation. Trees extract groundwater through their roots and release it into the atmosphere. When part of a forest is removed, the trees no longer transpire this water, resulting in a much drier climate. Deforestation reduces the content of water in the soil and groundwater as well as atmospheric moisture. The dry soil leads to lower water intake for the trees to extract. Deforestation reduces soil cohesion, so that erosion, flooding and landslides ensue.

Shrinking forest cover lessens the landscape's capacity to intercept, retain and transpire precipitation. Instead of trapping precipitation, which then percolates to groundwater systems, deforested areas become sources of surface water runoff, which moves much faster than subsurface flows. Forests return most of the water that falls as precipitation to the atmosphere by transpiration. In contrast, when an area is deforested, almost all precipitation is lost as run-off. That quicker transport of surface water can translate into flash flooding and more localized floods than would occur with the forest cover. Deforestation also contributes to decreased evapotranspiration, which lessens atmospheric moisture which in some cases affects precipitation levels downwind from the deforested area, as water is not recycled to downwind forests, but is lost in runoff and returns directly to the oceans. According to one study, in deforested north and northwest China, the average annual precipitation decreased by one third between the 1950s and the 1980s.

Trees, and plants in general, affect the water cycle significantly:

- their canopies intercept a proportion of precipitation, which is then evaporated back to the atmosphere (canopy interception);
- their litter, stems and trunks slow down surface runoff;
- their roots create macropores – large conduits – in the soil that increase infiltration of water;
- they contribute to terrestrial evaporation and reduce soil moisture via transpiration;

- their litter and other organic residue change soil properties that affect the capacity of soil to store water.
- their leaves control the humidity of the atmosphere by transpiring. 99% of the water absorbed by the roots moves up to the leaves and is transpired.

As a result, the presence or absence of trees can change the quantity of water on the surface, in the soil or groundwater, or in the atmosphere. This in turn changes erosion rates and the availability of water for either ecosystem functions or human services.

The forest may have little impact on flooding in the case of large rainfall events, which overwhelm the storage capacity of forest soil if the soils are at or close to saturation.

Tropical rainforests produce about 30% of our planet's fresh water.

Soil

Undisturbed forests have a very low rate of soil loss (erosion), approximately 2 metric tons per square kilometer (6 short tons per square mile). Deforestation generally increases rates of soil loss, by increasing the amount of runoff and reducing the protection of the soil from tree litter. This can be an advantage in excessively leached tropical rain forest soils. Forestry operations themselves also increase erosion through the development of (forest) roads and the use of mechanized equipment.

China's Loess Plateau was cleared of forest millennia ago. Since then it has been eroding, creating dramatic incised valleys, and providing the sediment that gives the Yellow River its yellow color and that causes the flooding of the river in the lower reaches (hence the river's nickname 'China's sorrow').

Removal of trees does not always increase erosion rates. In certain regions of southwest US, shrubs and trees have been encroaching on grassland. The trees themselves enhance the loss of grass between tree canopies. The bare intercanopy areas become highly erodible. The US Forest Service, in Bandelier National Monument for example, is studying how to restore the former ecosystem, and reduce erosion, by removing the trees.

Tree roots bind soil together, and if the soil is sufficiently shallow they act to keep the soil in place by also binding with underlying bedrock. Tree removal on steep slopes with shallow soil thus increases the risk of landslides, which can threaten people living nearby.

Biodiversity

Deforestation on a human scale results in decline in biodiversity, and on a natural global scale is known to cause the extinction of many species. The removal or destruction of areas of forest cover has resulted in a degraded environment with reduced biodiversity. Forests support biodiversity, providing habitat for wildlife; moreover, forests foster medicinal conservation. With forest biotopes being irreplaceable source of new drugs (such as taxol), deforestation can destroy genetic variations (such as crop resistance) irretrievably.

Illegal logging in Madagascar. In 2009, the vast majority of the illegally obtained rosewood was exported to China.

Since the tropical rainforests are the most diverse ecosystems on Earth and about 80% of the world's known biodiversity could be found in tropical rainforests, removal or destruction of significant areas of forest cover has resulted in a degraded environment with reduced biodiversity. A study in Rondônia, Brazil, has shown that deforestation also removes the microbial community which is involved in the recycling of nutrients, the production of clean water and the removal of pollutants.

It has been estimated that we are losing 137 plant, animal and insect species every single day due to rainforest deforestation, which equates to 50,000 species a year. Others state that tropical rainforest deforestation is contributing to the ongoing Holocene mass extinction. The known extinction rates from deforestation rates are very low, approximately 1 species per year from mammals and birds which extrapolates to approximately 23,000 species per year for all species. Predictions have been made that more than 40% of the animal and plant species in Southeast Asia could be wiped out in the 21st century. Such predictions were called into question by 1995 data that show that within regions of Southeast Asia much of the original forest has been converted to monospecific plantations, but that potentially endangered species are few and tree flora remains widespread and stable.

Scientific understanding of the process of extinction is insufficient to accurately make predictions about the impact of deforestation on biodiversity. Most predictions of forestry related biodiversity loss are based on species-area models, with an underlying assumption that as the forest declines species diversity will decline similarly. However, many such models have been proven to be wrong and loss of habitat does not necessarily lead to large scale loss of species. Species-area models are known to overpredict the number of species known to be threatened in areas where actual deforestation is ongoing, and greatly overpredict the number of threatened species that are widespread.

A recent study of the Brazilian Amazon predicts that despite a lack of extinctions thus far, up to 90 percent of predicted extinctions will finally occur in the next 40 years.

Economic Impact

Damage to forests and other aspects of nature could halve living standards for the world's poor and reduce global GDP by about 7% by 2050, a report concluded at the Convention on Biological Diversity (CBD) meeting in Bonn in 2008. Historically, utilization of forest products, including timber and fuel wood, has played a key role in human societies, comparable to the roles of water and cultivable land. Today, developed countries continue to utilize timber for building houses, and wood pulp for paper. In developing countries almost three billion people rely on wood for heating and cooking.

The forest products industry is a large part of the economy in both developed and developing countries. Short-term economic gains made by conversion of forest to agriculture, or over-exploitation of wood products, typically leads to loss of long-term income and long-term biological productivity. West Africa, Madagascar, Southeast Asia and many other regions have experienced lower revenue because of declining timber harvests. Illegal logging causes billions of dollars of losses to national economies annually.

The new procedures to get amounts of wood are causing more harm to the economy and overpower the amount of money spent by people employed in logging. According to a study, "in most areas studied, the various ventures that prompted deforestation rarely generated more than US$5 for every ton of carbon they released and frequently returned far less than US$1". The price on the European market for an offset tied to a one-ton reduction in carbon is 23 euro (about US$35).

Rapidly growing economies also have an effect on deforestation. Most pressure will come from the world's developing countries, which have the fastest-growing populations and most rapid economic (industrial) growth. In 1995, economic growth in developing countries reached nearly 6%, compared with the 2% growth rate for developed countries." As our human population grows, new homes, communities, and expansions of cities will occur. Connecting all of the new expansions will be roads, a very important part in our daily life. Rural roads promote economic development but also facilitate deforestation. About 90% of the deforestation has occurred within 100 km of roads in most parts of the Amazon.

The European Union is one of the largest importer of products made from illegal deforestation.

Forest Transition Theory

The forest area change may follow a pattern suggested by the forest transition (FT) theory, whereby at early stages in its development a country is characterized by high forest cover and low deforestation rates (HFLD countries).

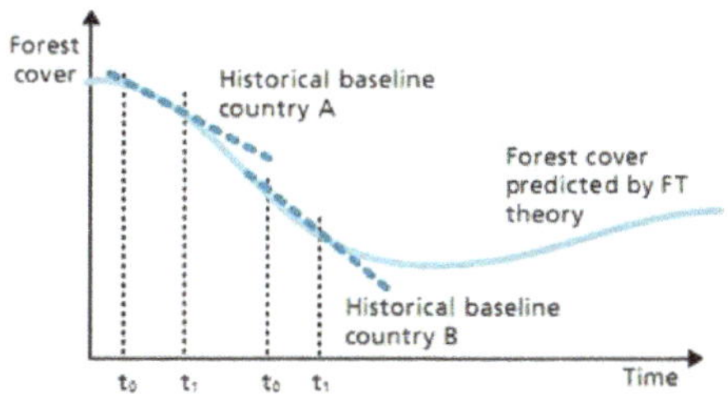

Source: Angelsen 2008.

The forest transition and historical baselines.

Then deforestation rates accelerate (HFHD, high forest cover – high deforestation rate), and forest cover is reduced (LFHD, low forest cover – high deforestation rate), before the deforestation rate slows (LFLD, low forest cover – low deforestation rate), after which forest cover stabilizes and eventually starts recovering. FT is not a "law of nature," and the pattern is influenced by national context (for example, human population density, stage of development, structure of the economy), global economic forces, and government policies. A country may reach very low levels of forest cover before it stabilizes, or it might through good policies be able to "bridge" the forest transition.

FT depicts a broad trend, and an extrapolation of historical rates therefore tends to underestimate future BAU deforestation for counties at the early stages in the transition (HFLD), while it tends to overestimate BAU deforestation for countries at the later stages (LFHD and LFLD).

Countries with high forest cover can be expected to be at early stages of the FT. GDP per capita captures the stage in a country's economic development, which is linked to the pattern of natural resource use, including forests. The choice of forest cover and GDP per capita also fits well with the two key scenarios in the FT:

(i) a forest scarcity path, where forest scarcity triggers forces (for example, higher prices of forest products) that lead to forest cover stabilization; and

(ii) an economic development path, where new and better off-farm employment opportunities associated with economic growth (= increasing GDP per capita) reduce profitability of frontier agriculture and slows deforestation.

Historical Causes

Prehistory

The Carboniferous Rainforest Collapse was an event that occurred 300 million years ago. Climate change devastated tropical rainforests causing the extinction of many plant and animal species. The change was abrupt, specifically, at this time climate became cooler and drier, conditions that are not favourable to the growth of rainforests and much of the biodiversity within them. Rainforests were fragmented forming shrinking 'islands' further and further apart. This sudden collapse affected several large groups, effects on amphibians were particularly devastating, while reptiles fared better, being ecologically adapted to the drier conditions that followed.

An array of Neolithic artifacts, including bracelets, axe heads, chisels, and polishing tools.

Rainforests once covered 14% of the earth's land surface; now they cover a mere 6% and experts estimate that the last remaining rainforests could be consumed in less than 40 years. Small scale deforestation was practiced by some societies for tens of thousands of years before the beginnings of civilization. The first evidence of deforestation appears in the Mesolithic period. It was probably used to convert closed forests into more open ecosystems favourable to game animals. With the advent of agriculture, larger areas began to be deforested, and fire became the prime tool to clear land for crops. In Europe there is little solid evidence before 7000 BC. Mesolithic foragers used fire to create openings for red deer and wild boar. In Great Britain, shade-tolerant species such as oak and ash are replaced in the pollen record by hazels, brambles, grasses and nettles. Removal of the forests led to decreased transpiration, resulting in the formation of upland peat bogs. Widespread decrease in elm pollen across Europe between 8400–8300 BC and 7200–7000 BC, starting in southern Europe and gradually moving north to Great Britain, may represent land clearing by fire at the onset of Neolithic agriculture.

The Neolithic period saw extensive deforestation for farming land. Stone axes were being made from about 3000 BC not just from flint, but from a wide variety of hard rocks from across Britain and North America as well. They include the noted Langdale axe industry in the English Lake District, quarries developed at Penmaenmawr in North Wales and numerous other locations. Rough-outs were made locally near the quarries, and some were polished locally to give a fine finish. This step not only increased the mechanical strength of the axe, but also made penetration of wood easier. Flint was still used from sources such as Grimes Graves but from many other mines across Europe.

Evidence of deforestation has been found in Minoan Crete; for example the environs of the Palace of Knossos were severely deforested in the Bronze Age.

Pre-industrial History

Throughout most of history, humans were hunter gatherers who hunted within forests. In most areas, such as the Amazon, the tropics, Central America, and the Caribbean, only after shortages of wood and other forest products occur are policies implemented to ensure forest resources are used in a sustainable manner.

In ancient Greece, Tjeered van Andel and co-writers summarized three regional studies of historic erosion and alluviation and found that, wherever adequate evidence exists, a major phase of erosion follows, by about 500-1,000 years the introduction of farming in the various regions of Greece, ranging from the later Neolithic to the Early Bronze Age. The thousand years following the mid-first millennium BC saw serious, intermittent pulses of soil erosion in numerous places. The historic silting of ports along the southern coasts of Asia Minor (*e.g.* Clarus, and the examples of Ephesus, Priene and Miletus, where harbors had to be abandoned because of the silt deposited by the Meander) and in coastal Syria during the last centuries BC.

Easter Island

Easter Island has suffered from heavy soil erosion in recent centuries, aggravated by agriculture and deforestation. Jared Diamond gives an extensive look into the collapse of the ancient Easter Islanders in his book *Collapse*. The disappearance of the island's trees seems to coincide with a decline of its civilization around the 17th and 18th century. He attributed the collapse to deforestation and over-exploitation of all resources.

The famous silting up of the harbor for Bruges, which moved port commerce to Antwerp, also followed a period of increased settlement growth (and apparently of deforestation) in the upper river basins. In early medieval Riez in upper Provence, alluvial silt from two small rivers raised the riverbeds and widened the floodplain, which slowly buried the Roman settlement in alluvium and gradually moved new construction to higher ground; concurrently the headwater valleys above Riez were being opened to pasturage.

A typical progress trap was that cities were often built in a forested area, which would provide wood for some industry (for example, construction, shipbuilding, pottery). When deforestation occurs without proper replanting, however; local wood supplies become difficult to obtain near enough to remain competitive, leading to the city's abandonment, as happened repeatedly in Ancient Asia Minor. Because of fuel needs, mining and metallurgy often led to deforestation and city abandonment.

With most of the population remaining active in (or indirectly dependent on) the agricultural sector, the main pressure in most areas remained land clearing for crop and cattle farming. Enough wild green was usually left standing (and partially used, for example, to collect firewood, timber and fruits, or to graze pigs) for wildlife to remain viable. The elite's (nobility and higher clergy) protection of their own hunting privileges and game often protected significant woodlands.

Major parts in the spread (and thus more durable growth) of the population were played by monastical 'pioneering' (especially by the Benedictine and Commercial orders) and some feudal lords' recruiting farmers to settle (and become tax payers) by offering relatively good legal and fiscal conditions. Even when speculators sought to encourage towns, settlers needed an agricultural belt around or sometimes within defensive walls. When populations were quickly decreased by causes such as the Black Death or devastating warfare (for example, Genghis Khan's Mongol hordes in eastern and central Europe, Thirty Years' War in Germany), this could lead to settlements being abandoned. The land was reclaimed by nature, but the secondary forests usually lacked the original biodiversity.

Deforestation of Brazil's Atlantic Forest c.1820-1825

From 1100 to 1500 AD, significant deforestation took place in Western Europe as a result of the expanding human population. The large-scale building of wooden sailing ships by European (coastal) naval owners since the 15th century for exploration, colonisation, slave trade–and other trade on the high seas consumed many forest resources. Piracy also contributed to the over harvesting of forests, as in Spain. This led to a weakening of the domestic economy after Columbus' discovery of America, as the economy became dependent on colonial activities (plundering, mining, cattle, plantations, trade, etc.)

In *Changes in the Land* (1983), William Cronon analyzed and documented 17th-century English colonists' reports of increased seasonal flooding in New England during the period when new settlers initially cleared the forests for agriculture. They believed flooding was linked to widespread forest clearing upstream.

The massive use of charcoal on an industrial scale in Early Modern Europe was a new type of consumption of western forests; even in Stuart England, the relatively primitive production of charcoal has already reached an impressive level. Stuart England was so widely deforested that it depended on the Baltic trade for ship timbers, and looked to the untapped forests of New England to supply the need. Each of Nelson's Royal Navy war ships at Trafalgar (1805) required 6,000 mature oaks for its construction. In France, Colbert planted oak forests to supply the French navy in the future. When the oak plantations matured in the mid-19th century, the masts were no longer required because shipping had changed.

Norman F. Cantor's summary of the effects of late medieval deforestation applies equally well to Early Modern Europe:

Europeans had lived in the midst of vast forests throughout the earlier medieval centuries. After 1250 they became so skilled at deforestation that by 1500 they were running short of wood for heating and cooking. They were faced with a nutritional decline because of the elimination of the generous supply of wild game that had inhabited the now-disappearing forests, which throughout medieval times had provided the staple of their carnivorous high-protein diet. By 1500 Europe was on the edge of a fuel and nutritional disaster [from] which it was saved in the sixteenth century only by the burning of soft coal and the cultivation of potatoes and maize.

Industrial Era

In the 19th century, introduction of steamboats in the United States was the cause of deforestation of banks of major rivers, such as the Mississippi River, with increased and more severe flooding one of the environmental results. The steamboat crews cut wood every day from the riverbanks to fuel the steam engines. Between St. Louis and the confluence with the Ohio River to the south, the Mississippi became more wide and shallow, and changed its channel laterally. Attempts to improve navigation by the use of snag pullers often resulted in crews' clearing large trees 100 to 200 feet (61 m) back from the banks. Several French colonial towns of the Illinois Country, such as Kaskaskia, Cahokia and St. Philippe, Illinois were flooded and abandoned in the late 19th century, with a loss to the cultural record of their archeology.

The wholescale clearance of woodland to create agricultural land can be seen in many parts of the world, such as the Central forest-grasslands transition and other areas of the Great Plains of the United States. Specific parallels are seen in the 20th-century deforestation occurring in many developing nations.

Rates of Deforestation

Global deforestation sharply accelerated around 1852. It has been estimated that about half of the Earth's mature tropical forests—between 7.5 million and 8 million km^2 (2.9 million to 3 million sq mi) of the original 15 million to 16 million km^2 (5.8 million to 6.2 million sq mi) that until 1947 covered the planet—have now been destroyed. Some scientists have predicted that unless significant measures (such as seeking out and protecting old growth forests that have not been disturbed) are taken on a worldwide basis, by 2030 there will only be 10% remaining, with another 10% in a degraded condition. 80% will have been lost, and with them hundreds of thousands of irreplaceable species. Some cartographers have attempted to illustrate the sheer scale of deforestation by country using a cartogram.

Slash-and-burn farming in the state of Rondônia, western Brazil

Estimates vary widely as to the extent of tropical deforestation. Scientists estimate that one fifth of the world's tropical rainforest was destroyed between 1960 and 1990. They claim that that rainforests 60 years ago covered 14% of the world's land surface, now only cover 5–7%, and that all tropical forests will be gone by the middle of the 21st century.

A 2002 analysis of satellite imagery suggested that the rate of deforestation in the humid tropics (approximately 5.8 million hectares per year) was roughly 23% lower than the most commonly quoted rates. Conversely, a newer analysis of satellite images reveals that deforestation of the Amazon rainforest is twice as fast as scientists previously estimated.

Some have argued that deforestation trends may follow a Kuznets curve, which if true would nonetheless fail to eliminate the risk of irreversible loss of non-economic forest values (for example, the extinction of species).

Satellite image of Haiti's border with the Dominican Republic (right) shows the amount of deforestation on the Haitian side

A 2005 report by the United Nations Food and Agriculture Organization (FAO) estimates that although the Earth's total forest area continues to decrease at about 13 million hectares per year, the global rate of deforestation has recently been slowing. Still others claim that rainforests are being destroyed at an ever-quickening pace. The London-based Rainforest Foundation notes that "the UN figure is based on a definition of forest as being an area with as little as 10% actual tree cover, which would therefore include areas that are actually savannah-like ecosystems and badly damaged forests." Other critics of the FAO data point out that they do not distinguish between forest types, and that they are based largely on reporting from forestry departments of individual countries, which do not take into account unofficial activities like illegal logging.

Despite these uncertainties, there is agreement that destruction of rainforests remains a significant environmental problem. Up to 90% of West Africa's coastal rainforests have disappeared since 1900. In South Asia, about 88% of the rainforests have been lost. Much of what remains of the world's rainforests is in the Amazon basin, where the Amazon Rainforest covers approximately 4 million square kilometres. The regions with the highest tropical deforestation rate between 2000 and 2005 were Central America—which lost 1.3% of its forests each year—and tropical Asia. In Central America, two-thirds of lowland tropical forests have been turned into pasture since 1950 and 40% of all the rainforests have been lost in the last 40 years. Brazil has lost 90–95% of its Mata Atlântica forest. Paraguay was losing its natural semi humid forests in the country's western regions at a rate of 15.000 hectares at a randomly studied 2-month period in 2010, Paraguay's parliament refused in 2009 to pass a law that would have stopped cutting of natural forests altogether.

Deforestation around Pakke Tiger Reserve, India

Madagascar has lost 90% of its eastern rainforests. As of 2007, less than 1% of Haiti's forests remained. Mexico, India, the Philippines, Indonesia, Thailand, Burma, Malaysia, Bangladesh, China, Sri Lanka, Laos, Nigeria, the Democratic Republic of the Congo, Liberia, Guinea, Ghana and the Ivory Coast, have lost large areas of their rainforest. Several countries, notably Brazil, have declared their deforestation a national emergency. The World Wildlife Fund's ecoregion project catalogues habitat types throughout the world, including habitat loss such as deforestation, showing for example that even in the rich forests of parts of Canada such as the Mid-Continental Canadian forests of the prairie provinces half of the forest cover has been lost or altered.

Regions

Rates of deforestation vary around the world.

In 2011 Conservation International listed the top 10 most endangered forests, characterized by having all lost 90% or more of their original habitat, and each harboring at least 1500 endemic plant species (species found nowhere else in the world).

Top 10 Most Endangered Forests 2011				
Endangered forest	Region	Remaining habitat	Predominate vegetation type	Notes
Indo-Burma	Asia-Pacific	5%	Tropical and subtropical moist broadleaf forests	Rivers, floodplain wetlands, mangrove forests. Burma, Thailand, Laos, Vietnam, Cambodia, India.
New Caledonia	Asia-Pacific	5%	Tropical and subtropical moist broadleaf forests	See note for region covered.
Sundaland	Asia-Pacific	7%	Tropical and subtropical moist broadleaf forests	Western half of the Indo-Malayan archipelago including southern Borneo and Sumatra.
Philippines	Asia-Pacific	7%	Tropical and subtropical moist broadleaf forests	Forests over the entire country including 7,100 islands.
Atlantic Forest	South America	8%	Tropical and subtropical moist broadleaf forests	Forests along Brazil's Atlantic coast, extends to parts of Paraguay, Argentina and Uruguay.
Mountains of Southwest China	Asia-Pacific	8%	Temperate coniferous forest	See note for region covered.
California Floristic Province	North America	10%	Tropical and subtropical dry broadleaf forests	See note for region covered.
Coastal Forests of Eastern Africa	Africa	10%	Tropical and subtropical moist broadleaf forests	Mozambique, Tanzania, Kenya, Somalia.
Madagascar & Indian Ocean Islands	Africa	10%	Tropical and subtropical moist broadleaf forests	Madagascar, Mauritius, Reunion, Seychelles, Comoros.
Eastern Afromontane	Africa	11%	Tropical and subtropical moist broadleaf forests Montane grasslands and shrublands	Forests scattered along the eastern edge of Africa, from Saudi Arabia in the north to Zimbabwe in the south.

Control

Reducing Emissions

Main international organizations including the United Nations and the World Bank, have begun to develop programs aimed at curbing deforestation. The blanket term Reducing Emissions from Deforestation and Forest Degradation (REDD) describes these sorts of programs, which use direct monetary or other incentives to encourage developing countries to limit and/or roll back deforestation. Funding has been an issue, but at the UN Framework Convention on Climate Change (UNFCCC) Conference of the Parties-15 (COP-15) in Copenhagen in December 2009, an accord was reached with a collective commitment by developed countries for new and additional resources, including forestry and investments through international institutions, that will approach USD 30 billion for the period 2010–2012. Significant work is underway on tools for use in monitoring developing country adherence to their agreed REDD targets. These tools, which rely on remote forest monitoring using satellite imagery and other data sources, include the Center for Global Development's FORMA (Forest Monitoring for Action) initiative and the Group on Earth Observations' Forest Carbon Tracking Portal. Methodological guidance for forest monitoring was also emphasized at COP-15. The environmental organization Avoided Deforestation Partners leads the campaign for development of REDD through funding from the U.S. government. In 2014, the Food and Agriculture Organization of the United Nations and partners launched Open Foris - a set of open-source software tools that assist countries in gathering, producing and disseminating information on the state of forest resources. The tools support the inventory lifecycle, from needs assessment, design, planning, field data collection and management, estimation analysis, and dissemination. Remote sensing image processing tools are included, as well as tools for international reporting for Reducing emissions from deforestation and forest degradation (REDD) and MRV and FAO's Global Forest Resource Assessments.

In evaluating implications of overall emissions reductions, countries of greatest concern are those categorized as High Forest Cover with High Rates of Deforestation (HFHD) and Low Forest Cover with High Rates of Deforestation (LFHD). Afghanistan, Benin, Botswana, Burma, Burundi, Cameroon, Chad, Ecuador, El Salvador, Ethiopia, Ghana, Guatemala, Guinea, Haiti, Honduras, Indonesia, Liberia, Malawi, Mali, Mauritania, Mongolia, Namibia, Nepal, Nicaragua, Niger, Nigeria, Pakistan, Paraguay, Philippines, Senegal, Sierra Leone, Sri Lanka, Sudan, Togo, Uganda, United Republic of Tanzania, Zimbabwe are listed as having Low Forest Cover with High Rates of Deforestation (LFHD). Brazil, Cambodia, Democratic Peoples Republic of Korea, Equatorial Guinea, Malaysia, Solomon Islands, Timor-Leste, Venezuela, Zambia are listed as High Forest Cover with High Rates of Deforestation (HFHD).

Payments for Conserving Forests

In Bolivia, deforestation in upper river basins has caused environmental problems, including soil erosion and declining water quality. An innovative project to try and remedy this situation involves landholders in upstream areas being paid by downstream water users to conserve forests. The landholders receive US$20 to conserve the trees, avoid polluting livestock practices, and enhance the biodiversity and forest carbon on their land. They also receive US$30, which purchases a beehive, to compensate for conservation for two hectares of water-sustaining forest for five years. Honey revenue per hectare of forest is US$5 per year, so within five years, the landholder has sold US$50 of honey. The project is being conducted by Fundación Natura Bolivia and Rare Conservation, with support from the Climate & Development Knowledge Network.

Land Rights

Transferring land rights to indigenous inhabitants is argued to efficiently conserve forests.

Transferring rights over land from public domain to its indigenous inhabitants is argued to be a cost effective strategy to conserve forests. This includes the protection of such rights entitled in existing laws, such as India's Forest Rights Act. The transferring of such rights in China, perhaps the largest land reform in modern times, has been argued to have increased forest cover. In Brazil, forested areas given tenure to indigenous groups have even lower rates of clearing than national parks.

Farming

New methods are being developed to farm more intensively, such as high-yield hybrid crops, greenhouse, autonomous building gardens, and hydroponics. These methods are often dependent on chemical inputs to maintain necessary yields. In cyclic agriculture, cattle are grazed on farm land that is resting and rejuvenating. Cyclic agriculture actually increases the fertility of the soil. Intensive farming can also decrease soil nutrients by consuming at an accelerated rate the trace minerals needed for crop growth.The most promising approach, however, is the concept of food forests in permaculture, which consists of agroforestal systems carefully designed to mimic natural forests, with an emphasis on plant and animal species of interest for food, timber and other uses. These systems have low dependence on fossil fuels and agro-chemicals, are highly self-maintaining, highly productive, and with strong positive impact on soil and water quality, and biodiversity.

Monitoring Deforestation

There are multiple methods that are appropriate and reliable for reducing and monitoring deforestation. One method is the "visual interpretation of aerial photos or satellite imagery that is labor-intensive but does not require high-level training in computer image processing or extensive computational resources". Another method includes hot-spot analysis (that is, locations of rapid change) using expert opinion or coarse resolution satellite data to identify locations for detailed digital analysis with high resolution satellite images. Deforestation is typically assessed by quantifying the amount of area deforested, measured at the present time. From an environmental point of view, quantifying the damage and its possible consequences is a more important task, while conservation efforts are more focused on forested land protection and development of land-use alternatives to avoid continued deforestation. Deforestation rate and total area deforested, have been widely used for monitoring deforestation in many regions, including the Brazilian Amazon deforestation monitoring by INPE.

Forest Management

Efforts to stop or slow deforestation have been attempted for many centuries because it has long been known that deforestation can cause environmental damage sufficient in some cases to cause societies to collapse. In Tonga, paramount rulers developed policies designed to prevent conflicts between short-term gains from converting forest to farmland and long-term problems forest loss would cause, while during the 17th and 18th centuries in Tokugawa, Japan, the shoguns developed a highly sophisticated system of long-term planning to stop and even reverse deforestation of the preceding centuries through substituting timber by other products and more efficient use of land that had been farmed for many centuries. In 16th-century Germany, landowners also developed silviculture to deal with the problem of deforestation. However, these policies tend to be limited to environments with *good rainfall, no dry season* and *very young soils* (through volcanism or glaciation). This is because on older and less fertile soils trees grow too slowly for silviculture to be economic, whilst in areas with a strong dry season there is always a risk of forest fires destroying a tree crop before it matures.

In the areas where "slash-and-burn" is practiced, switching to "slash-and-char" would prevent the rapid deforestation and subsequent degradation of soils. The biochar thus created, given back to the soil, is not only a durable carbon sequestration method, but it also is an extremely beneficial amendment to the soil. Mixed with biomass it brings the creation of terra preta, one of the richest soils on the planet and the only one known to regenerate itself.

Sustainable Practices

Certification, as provided by global certification systems such as Programme for the Endorsement of Forest Certification and Forest Stewardship Council, contributes to tackling deforestation by creating market demand for timber from sustainably managed forests. According to the United Nations Food and Agriculture Organization (FAO), "A major condition for the adoption of sustainable forest management is a demand for products that are produced sustainably and consumer willingness to pay for the higher costs entailed. Certification represents a shift from regulatory approaches to market incentives to promote sustainable forest management. By promoting the positive attributes of forest products from sustainably managed forests, certification focuses on the demand side of environmental conservation." Rainforest Rescue argues that the standards of organizations like FSC are too closely connected to timber industry interests and therefore do not guarantee environmentally and socially responsible forest management. In reality, monitoring systems are inadequate and various cases of fraud have been documented worldwide.

Bamboo is advocated as a more sustainable alternative for cutting down wood for fuel.

Some nations have taken steps to help increase the amount of trees on Earth. In 1981, China created National Tree Planting Day Forest and forest coverage had now reached 16.55% of China's land mass, as against only 12% two decades ago

Using fuel from bamboo rather than wood results in cleaner burning, and since bamboo matures much faster than wood, deforestation is reduced as supply can be replenished faster.

Reforestation

In many parts of the world, especially in East Asian countries, reforestation and afforestation are increasing the area of forested lands. The amount of woodland has increased in 22 of the world's 50 most forested nations. Asia as a whole gained 1 million hectares of forest between 2000 and 2005. Tropical forest in El Salvador expanded more than 20% between 1992 and 2001. Based on these trends, one study projects that global forest will increase by 10%—an area the size of India—by 2050.

In the People's Republic of China, where large scale destruction of forests has occurred, the government has in the past required that every able-bodied citizen between the ages of 11 and 60 plant three to five trees per year or do the equivalent amount of work in other forest services. The government claims that at least 1 billion trees have been planted in China every year since 1982. This is no longer required today, but March 12 of every year in China is the Planting Holiday. Also, it has introduced the Green Wall of China project, which aims to halt the expansion of the Gobi desert through the planting of trees. However, due to the large percentage of trees dying off after planting (up to 75%), the project is not very successful. There has been a 47-million-hectare increase in forest area in China since the 1970s. The total number of trees amounted to be about 35 billion and 4.55% of China's land mass increased in forest coverage. The forest coverage was 12% two decades ago and now is 16.55%.

An ambitious proposal for China is the Aerially Delivered Re-forestation and Erosion Control System and the proposed Sahara Forest Project coupled with the Seawater Greenhouse.

In Western countries, increasing consumer demand for wood products that have been produced and harvested in a sustainable manner is causing forest landowners and forest industries to become increasingly accountable for their forest management and timber harvesting practices.

The Arbor Day Foundation's Rain Forest Rescue program is a charity that helps to prevent deforestation. The charity uses donated money to buy up and preserve rainforest land before the lumber companies can buy it. The Arbor Day Foundation then protects the land from deforestation. This also locks in the way of life of the primitive tribes living on the forest land. Organizations such as Community Forestry International, Cool Earth, The Nature Conservancy, World Wide Fund for Nature, Conservation International, African Conservation Foundation and Greenpeace also focus on preserving forest habitats. Greenpeace in particular has also mapped out the forests that are still intact and published this information on the internet. World Resources Institute in turn has made a simpler thematic map showing the amount of forests present just before the age of man (8000 years ago) and the current (reduced) levels of forest. These maps mark the amount of afforestation required to repair the damage caused by people.

Forest Plantations

To meet the world's demand for wood, it has been suggested by forestry writers Botkins and Sedjo that high-yielding forest plantations are suitable. It has been calculated that plantations yielding 10 cubic meters per hectare annually could supply all the timber required for international trade on 5% of the world's existing forestland. By contrast, natural forests produce about 1–2 cubic meters per hectare; therefore, 5–10 times more forestland would be required to meet demand. Forester Chad Oliver has suggested a forest mosaic with high-yield forest lands interspersed with conservation land.

In the country of Senegal, on the western coast of Africa, a movement headed by youths has helped to plant over 6 million mangrove trees. The trees will protect local villages from storm damages and will provide a habitat for local wildlife. The project started in 2008, and already the Senegalese government has been asked to establish rules and regulations that would protect the new mangrove forests.

Military Context

American Sherman tanks knocked out by Japanese artillery on Okinawa.

While the preponderance of deforestation is due to demands for agricultural and urban use for the human population, there are some examples of military causes. One example of deliberate deforestation is that which took place in the U.S. zone of occupation in Germany after World War II. Before the onset of the Cold War, defeated Germany was still considered a potential future threat rather than potential future ally. To address this threat, attempts were made to lower German industrial potential, of which forests were deemed an element. Sources in the U.S. government admitted that the purpose of this was that the "ultimate destruction of the war potential of German forests." As a consequence of the practice of clear-felling, deforestation resulted which could "be replaced only by long forestry development over perhaps a century."

Deforestation can also be one consequence of war. For example, in the 1945 Battle of Okinawa, bombardment and other combat operations reduced the lush tropical landscape into "a vast field of mud, lead, decay and maggots". Deforestation can also be an intentional tactic of military forces. Defoliants (like Agent Orange or others) was used by the British in the Malayan Emergency, and by the United States in the Korean War and Vietnam War.

Public Health Context

Deforestation eliminates a great number of species of plants and animals which also often results in an increase in disease. Loss of native species allows new species to come to dominance. Often the destruction of predatory species can result in an increase in rodent populations. These are known to carry plagues. Additionally, erosion can produce pools of stagnant water that are perfect breeding grounds for mosquitos, well known vectors of malaria, yellow fever, nipah virus, and more. Deforestation can also create a path for non-native species to flourish such as certain types of snails, which have been correlated with an increase in schistosomiasis cases.

Deforestation is occurring all over the world and has been coupled with an increase in the occurrence of disease outbreaks. In Malaysia, thousands of acres of forest have been cleared for pig farms. This has resulted in an increase in the zoonosis the Nipah virus. In Kenya, deforestation has led to an increase in malaria cases which is now the leading cause of morbidity and mortality the country.

Another pathway through which deforestation affects disease is the relocation and dispersion of disease-carrying hosts. This disease emergence pathway can be called "range expansion," whereby the host's range (and thereby the range of pathogens) expands to new geographic areas. Through deforestation, hosts and reservoir species are forced into neighboring habitats. Accompanying the reservoir species are pathogens that have the ability to find new hosts in previously unexposed regions. As these pathogens and species come into closer contact with humans, they are infected both directly and indirectly.

A catastrophic example of range expansion is the 1998 outbreak of Nipah Virus in Malaysia. For a number of years, deforestation, drought, and subsequent fires led to a dramatic geographic shift and density of fruit bats, a reservoir for Nipah virus. Deforestation reduced the available fruiting trees in the bats' habitat, and they encroached on surrounding orchards which also happened to be the location of a large number of pigsties. The bats, through proximity spread the Nipah to pigs. While the virus infected the pigs, mortality was much lower than among humans, making the pigs a virulent host leading to the transmission of the virus to humans. This resulted in 265 reported cases of encephalitis, of which 105 resulted in death. This example provides an important lesson for the impact deforestation can have on human health.

Another example of range expansion due to deforestation and other anthropogenic habitat impacts includes the Capybara rodent in Paraguay. This rodent is the host of a number of zoonotic diseases and, while there has not yet been a human-borne outbreak due to the movement of this rodent into new regions, it offers an example of how habitat destruction through deforestation and subsequent movements of species is occurring regularly.

A now well-developed theory is that the spread of HIV it is at least partially due deforestation. Rising populations created a food demand and with deforestation opening up new areas of the forest the hunters harvested a great deal of primate bushmeat, which is believed to be the origin of HIV.

Illegal Logging

Illegal logging is the harvest, transportation, purchase or sale of timber in violation of laws. The harvesting procedure itself may be illegal, including using corrupt means to gain access to forests; extraction without permission or from a protected area; the cutting of protected species; or the extraction of timber in excess of agreed limits.

Illegalities may also occur during transport, such as illegal processing and export; fraudulent declaration to customs; the avoidance of taxes and other charges, and fraudulent certification.

Overview

Illegal logging is a pervasive problem, causing enormous damage to forests, local communities and to the economies of producer countries. Despite the economic importance of trade in timber and forest products, major international timber consumer countries, such as the EU, have no legal means to halt the import of illegally sourced forest products, because the identification of illegally logged or traded timber is technically difficult. Therefore, a legal basis for normative acts against timber imports or other products manufactured out of illegal wood is missing. Scientific methods to pinpoint the geographic origin of timber are currently under development. Possible actions to restrict imports cannot meet with WTO regulations of non-discrimination. They must instead be arranged in bilateral agreements. TRAFFIC, the wildlife trade monitoring network strives to monitor the illegal trade of timber and provide expertise in policy and legal reviews.

Scale

It is estimated that illegal logging in public lands alone causes losses in assets and revenue in excess of 10 billion USD annually. Although exact figures are difficult to state, given the illegal nature of the activity, decant estimates show that most of the logging, more than half, that is done in the world is illegal most especially in open and vulnerable areas – the Amazon Basin, Central Africa, Southeast Asia, the Russian Federation – is illegal.

Available figures and estimations must be dealt with cautiously. Governments may tend to underestimate the situation. High estimates of illegal logging may constitute an embarrassment as these hint at ineffective enforcement of legislation or, even worse, bribery and corruption. On the other hand, environmental NGOs publish alarming figures to raise awareness and to emphasise the need for stricter conservation measures. For companies of the forest sector, publications providing high estimations can be regarded as potentially threatening for their reputation and their market perspective, including the competitiveness of wood in comparison to other materials. However, for many countries, NGOs are the only source of information apart from state institutions that probably clearly underestimate the situation. For example, the Republic of Estonia calculated an amount of 1% illegally harvested timber in 2003, whereas it is estimated to reach a maximum of 50% by the ENGO "Estonian Green Movement". In Latvia, the situation is comparable; anecdotal evidence points towards 25% of logging being illegal.

Consequences

Illegal logging contributes to deforestation and by extension global warming, causes loss of biodiversity and undermines the rule of law. These illegal activities undermine responsible forest management, encourage corruption and tax evasion and reduce the income of the producer countries, further limiting the resources producer countries can invest in sustainable development. Illegal logging has serious economic and social implications for the poor and disadvantaged with millions of dollars worth of timber revenue being lost each year.

Illegal logging continues in Thailand. This photograph was taken from the roadside in Mae Wang District, Chiang Mai Province, in March 2011

Furthermore, the illegal trade of forest resources undermines international security, and is frequently associated with corruption, money laundering, organized crime, human rights abuses and, in some cases, violent conflict. In the forestry sector, cheap imports of illegal timber and forest products, together with the non-compliance of some economic players with basic social and environmental standards, destabilise international markets. This unfair competition affects those European companies, especially the small and medium-sized companies that are behaving responsibly and ready to play by fair rules.

Statistics on Illegal Logging

- A joint UK-Indonesian study of the timber industry in Indonesia in 1998 suggested that about 40% of throughput was illegal, with a value in excess of $365 million. More recent estimates, comparing legal harvesting against known domestic consumption plus exports, suggest that 88% of logging in the country is illegal in some way. Malaysia is the key transit country for illegal wood products from Indonesia.
- In Brazil, 80% of logging in the Amazon violates government controls. At the core of illegal logging is widespread corruption. Often referred to as 'green gold', mahogany can fetch over US$1,600 m-3. Illegal mahogany opens the door for illegal logging of other species, and for widespread exploitation of the Brazilian Amazon. Recent Greenpeace investigations in the Brazilian state of Pará reveal just how deeply rooted the problem remains. No reliable legal chain of custody exists for mahogany, and the key players in its trade are ruthless.
- The World Bank estimates that 80% of logging operations are illegal in Bolivia and 42% in Colombia, 10 while in Peru, illegal logging equals 80% of all activities.

- Research carried out by WWF International in 2002 shows that in Africa, rates of illegal logging vary from 50% for Cameroon and Equatorial Guinea to 70% in Gabon and 80% in Liberia – where revenues from the timber industry also fuelled the civil war.
- WWF estimates that illegal logging in Russia is at least 20%, reaching up to 50% in its far eastern regions.
- A 2012 joint study by the United Nations Environment Programme and Interpol states that illegal logging accounts for up to 30% of the global logging trade and contributes to more than 50% of tropical deforestation in Central Africa, the Amazon Basin and South East Asia.
- Between 50% and 90% of logging from the key countries in these regions is being carried out by organised criminal entities.

Political Processes

East Asia

Signs of illegal timber poaching on the boundary of the protected area around the Cagua Volcano, Cagayan, Philippines.

The East Asia Forest Law Enforcement and Governance (EA FLEG) Ministerial Conference took place in Bali in September 2001. The Conference brought together nearly 150 participants from 20 countries, representing government, international organizations, NGOs, and the private sector. The event was co-hosted by the World Bank and the Government of Indonesia.

The meeting included detailed technical discussions of forest law enforcement in relation to governance, forest policy and forest management as well as ministerial engagement.

The Conference's primary aims were to share analysis on forest law enforcement; explore priority issues of forest law enforcement, including illegal logging in the East Asia region, among senior officials from forest and related ministries, NGOs and industry representatives; and commit to action at the national and regional level.

European Union

In May 2003, the European Commission presented the EU Forest Law Enforcement, Governance and Trade Action Plan (EU FLEGT). This marked the beginning of a long process by which the EU aims to develop and implement measures to address illegal logging and related trade. The primary means of implementing the Plan is through Voluntary Partnership Agreements with timber producing countries. The EU Timber Regulation (EUTR) was adopted in 2010 and went into effect 3 March 2013.

- It prohibits the placing on the EU market for the first time of illegally harvested timber and products derived from such timber;
- It requires EU traders who place timber products on the EU market for the first time to exercise 'due diligence';
- Once on the market, the timber and timber products may be sold on and/or transformed before they reach the final consumer.
- To facilitate the traceability of timber products economic operators in this part of the supply chain (referred to as traders in the regulation) have an obligation to keep records of their suppliers and customers.

A Greenpeace investigation published in May 2014 demonstrates that EU Timber Regulation is ineffective if fraudulent paperwork is accepted at face value and there is not sufficient enforcement by EU authorities.

Africa

The *Africa Forest Law Enforcement and Governance* (AFLEG) Ministerial Conference was held in Yaoundé, Cameroon in October 2003. The meeting drew together ministers and stakeholders from Africa, Europe and North America to consider how partnerships between producers, consumers, donors, civil society and the private sector could address illegal forest exploitation and associated trade in Africa.

The AFLEG conference, the second regional forest law enforcement and governance meeting after East Asia, resulted in endorsement of a ministerial declaration and action plan as well as a variety of informal implementation initiatives.

In 2014, the EU-FAO Forest Law Enforcement, Governance and Trade Programme of the Food and Agriculture Organization of the United Nations published the study *The Voluntary Partnership Agreement (VPA) process in Central and West Africa: from theory to practice* to document and foster strategic reflection in partner countries already engaged in negotiating a VPA - or those who will be entering into such negotiations - by providing examples of good practices. These good practices were identified and recorded following interviews with the main stakeholders in the eight VPA countries in West and Central Africa, the European Forest Institute's (EFI) EU FLEGT Facility and the European Commission.

Saint Petersburg Declaration

The *Europe and North Asia Forest Law Enforcement and Governance* (ENA FLEG) Ministerial Conference was held in Saint Petersburg, Russia on 22–25 November 2005. In May 2004, the Russian Federation announced its intention to host the ENA FLEG process, supported by the World Bank. A preparatory conference was held in Moscow in June 2005.

The Saint Petersburg conference brought together nearly 300 participants representing 43 governments, the private sector, civil society and international organisations. It agreed to the *Saint Petersburg Declaration on Forest Law Enforcement and Governance in Europe and North Asia*. The Declaration includes an indicative list of actions, intended to serve as a general framework for possible actions to be undertaken by governments as well as civil society.

The conference took place as the United Kingdom prepared to pass the G8 Presidency to Russia. As Valery Roshchupkin, Head of the Federal Forestry Agency of the Russian Federation, confirmed, illegal logging would be of special importance for Russia as the G8 President and for the following G8 Summit, also held in Saint Petersburg.

United States

In response to growing concerns over illegal logging and advice from TRAFFIC and other organisations, on May 22, 2008 the U.S. amended the Lacey Act, when the Food, Conservation, and Energy Act of 2008 expanded its protection to a broader range of plants and plant products (Section 8204. Prevention of Illegal Logging Practices).

The requirements under the new Amendments are two-fold. First, the Lacey Act now makes it illegal to import into the United States plants that have been harvested contrary to any applicable Federal Law, State Law, Indian Tribal Law, or Foreign Law. If a plant is found to have been harvested in violation of the laws of the country where it was harvested, that plant would be subject to seizure and forfeiture if imported into the U.S. The Lacey Act also makes it unlawful, beginning December 15, 2008, to import certain plants and plant products without a Plant and Plant Product import declaration.

This Plant and Plant Product Declaration must contain (among other things) the Genus, Species, and Country of Harvest of every plant found in commercial shipments of certain products, a list of applicable products (along with other requirements and guidance) can be found on the USDA APHIS website.

Australia

The Timber Development Association (TDA) welcomes on June 6, 2014's release by the Australian Department of Agriculture of a position paper on the Illegal Logging Prohibition Regulation and guidance on how timber and wood products industry can comply on the Australian Government - Department of Agriculture official website. The release of the Government's guidance coincides with the release of industry developed timber due diligence tools and information through the industry website of Timber Due Diligence.

The Australian Illegal Logging Prohibition Regulation applies to importers into Australia of "regulated timber products" such as sawn timber, wood panels, pulp, paper products, and wood furniture. The Regulation starts on 30 November 2014 and requires that before import of these products or processing of raw logs, due diligence is undertaken to minimise the risk that the timber products or raw logs were illegally logged or incorporate illegally logged timber.

Deforestation and Climate Change

Deforestation is one of the main causes of climate change. It is the second largest anthropogenic source of carbon dioxide to the atmosphere, after fossil fuel combustion. Deforestation and forest degradation contribute to atmospheric greenhouse gas emissions through combustion of forest biomass and decomposition of remaining plant material and soil carbon. It used to account for more than 20% of carbon dioxide emissions, but it's currently somewhere around the 10% mark. By 2008, deforestation was 12% of total CO_2, or 15% if peatlands are included. These proportions are likely to have fallen since given the continued rise of fossil fuel use.

Averaged over all land and ocean surfaces, temperatures warmed roughly 1.53 °F (0.85 °C) between 1880 and 2012, according to the Intergovernmental Panel on Climate Change. In the Northern Hemisphere, 1983 to 2012 were the warmest 30-year period of the last 1400 years.

Effect on Climate Change

Decrease in Biodiversity

A 2007 study conducted by the National Science Foundation found that biodiversity and genetic diversity are codependent—that diversity among species requires diversity within a species, and vice versa. "If any one type is removed from the system, the cycle can break down, and the community becomes dominated by a single species."

Counteracting Climate Change

Reforestation

Reforestation is the natural or intentional restocking of existing forests and woodlands that have been depleted, usually through deforestation. It is the reestablishment of forest cover either naturally or artificially. Similar to the other methods of forestation, reforestation can be very effective because a single tree can absorb as much as 48 pounds of carbon dioxide per year and can sequester 1 ton of carbon dioxide by the time it reaches 40 years old.

Afforestation

Afforestation is the establishment of a forest or stand of trees in an area where there was no forest.

China

Although China has set official goals for reforestation, these goals were set for an 80-year time horizon and were not significantly met by 2008. China is trying to correct these problems with projects such as the Green Wall of China, which aims to replant forests and halt the expansion of the Gobi Desert. A law promulgated in 1981 requires that every school student over the age of 11 plant at least one tree per year. But average success rates, especially in state-sponsored plantings, remains relatively low. And even the properly planted trees have had great difficulty surviving the combined impacts of prolonged droughts, pest infestation and fires. Nonetheless, China currently has the highest afforestation rate of any country or region in the world, with 4.77 million hectares (47,000 square kilometers) of afforestation in 2008.

Japan

The primary goal of afforestation projects in Japan is to develop the forest structure of the nation and to maintain the biodiversity found in the Japanese wilderness. The Japanese temperate rainforest is scattered throughout the Japanese archipelago and is home to many endemic species that are not naturally found anywhere else. As development of the country's caused a decline in forest cover, a reduction in biodiversity was seen in those areas.

Agroforestry

Agroforestry or agro-sylviculture is a land use management system in which trees or shrubs are grown around or among crops or pastureland. It combines agricultural and forestry technologies to create more diverse, productive, profitable, healthy, and sustainable land-use systems.

Projects and Foundations

Arbor Day Foundation

Founded in 1972, the centennial of the first Arbor Day observance in the 19th century, the Foundation has grown to become the largest nonprofit membership organization dedicated to planting trees, with over one million members, supporters, and valued partners. They work on projects focused on planting trees around campuses, low-income communities, and communities that have been affected by natural disasters among other places.

Billion Tree Campaign

The Billion Tree Campaign was launched in 2006 by the United Nations Environment Programme (UNEP) as a response to the challenges of global warming, as well as to a wider array of sustainability challenges, from water supply to biodiversity loss. Its initial target was the planting of one billion trees in 2007. Only one year later in 2008, the campaign's objective was raised to 7 billion trees—a target to be met by the climate change conference that was held in Copenhagen, Denmark in December 2009. Three months before the conference, the 7 billion planted trees mark had been surpassed. In December 2011, after more than 12 billion trees had been planted, UNEP formally handed management of the program over to the not-for-profit Plant-for-the-Planet initiative, based in Munich, Germany.

The Amazon Fund (Brazil)

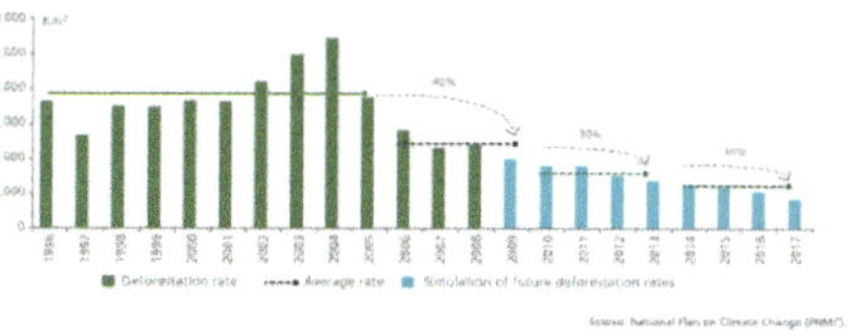

Four-year plan to reduce in deforestation in the Amazon

Considered the largest reserve of biological diversity in the world, the Amazon Basin is also the largest Brazilian biome, taking up almost half the nation's territory. The Amazon Basin corresponds to two fifths of South America's territory. Its area of approximately seven million square kilometers covers the largest hydrographic network on the planet, through which runs about one fifth of the fresh water on the world's surface. Deforestation in the Amazon rainforest is a major cause to climate change due to the decreasing number of trees available to capture increasing carbon dioxide levels in the atmosphere.

The Amazon Fund is aimed at raising donations for non-reimbursable investments in efforts to prevent, monitor and combat deforestation, as well as to promote the preservation and sustainable use of forests in the Amazon Biome, under the terms of Decree N.º 6,527, dated August 1, 2008. The Amazon Fund supports the following areas: management of public forests and protected areas, environmental control, monitoring and inspection, sustainable forest management, economic activities created with sustainable use of forests, ecological and economic zoning, territorial arrangement and agricultural regulation, preservation and sustainable use of biodiversity, and recovery of deforested areas. Besides those, the Amazon Fund may use up to 20% of its donations to support the development of systems to monitor and control deforestation in other Brazilian biomes and in biomes of other tropical countries.

Forest Dieback

Forest dieback (also "Waldsterben", a German loan word) is a condition in trees or woody plants in which peripheral parts are killed, either by pathogens, parasites or due to conditions like acid rain and drought. Two of the nine tipping points for major climate changes, forecast for the next century, are directly related to forest diebacks. *Phomopsis azadirachtae* is a fungus of the Phomopsis genus that has been identified as the fungus responsible for dieback of *Azadirachta indica* (Neem) in India.

Jizera Mountains in Central Europe

Definition

Forest dieback refers to the phenomena of a stand of trees losing health and dying without an obvious cause. This condition is also known as forest decline, forest damage, canopy level dieback, and stand level dieback (also Waldsterben and Waldschäden, German loan words). This usually affects individual species of trees, but can also affect multiple species. Trees and woody plants are killed by a complex group of factors that include parasites, such as fungus and beetles, pollution, such as acid rain and organic compounds, and drought. Diseases and pests can kill large groups of trees very easily, but the premature and progressive loss of clusters of trees without an obvious or specific cause is known as forest dieback. Dieback is an episodic event. Dieback takes on many locations and shapes. It can be along the perimeter, at specific elevations, or dispersed throughout the forest ecosystem.

Forest dieback has many symptoms: falling off of leaves and needles, discoloration of leaves and needles, thinning of the crowns of trees, dead stands of trees of a certain age, and changes in the roots of the trees. Forest dieback has many dynamic forms. A stand of trees can exhibit mild symptoms, extreme systems, or death at the same time. Forest decline can be viewed as the result of continued, widespread, and severe dieback of multiple species in the forest. Current forest decline is described by rapid development on individual trees, occurrence in different forest types, a long duration (over 10 years), and occurrence throughout the natural range of affected species.

History

Much research was done in the 1980s when a severe dieback happened in Germany and the Northeast United States. Previous diebacks were regionally limited, but starting at the end of the 1970s a decline took over the forests of Central Europe and parts of North

America. The forest damage in Germany was different because the decline was severe, up to 50% of the trees, prolonged, over 5 years, and widespread across tree species. Affected trees went from 8% in 1982 to 50% in 1984 and stayed at 50% through 1987. Many hypotheses have been proposed for this dieback.

Drought and temperature have been causal factors in many diebacks. This condition is common in semi-arid regions and the trees are already stressed by dehydration. There have been reported diebacks in the 50% mortality range across multiple species in Africa, Asia, Europe, North America, South and Central America, and Australia. Notable drought related diebacks have occurred globally from the 1970s through the 2000s. A few drought related diebacks have occurred before this. For instance, a 23% mortality rate was reported between 1945 and 1993 in Senegal Africa. Trees stressed by drought and temperature are more susceptible to parasites.

In North America there have been five notable hardwood diebacks this century. These have occurred with the maturation of the forest and each episode has lasted about eleven years. The most severe temperate forest dieback was on white birch and yellow trees. The others are ash, oak, and maple species. White and yellow birches experienced an episode starting between 1934 and 1957 and ending between 1953 and 1954. This followed a wave pattern that first appeared in southern regions and moved to northern regions. A second wave is evident between 1957 and 1965 in northern Quebec. Sugar maple experienced a wave of dieback in parts of the United States during the 1960s. A second wave occurred primarily in Canada during the 1980s, but also reached the United States. These diebacks were numerically analyzed to exclude natural tree mortality. It is hypothesized that a mature forest is more susceptible to extreme environmental stresses.

Pathogens are responsible for many diebacks. It is difficult to isolate and identify exactly which pathogens are responsible and how they interact with the trees. For instance *Phomopsis azadirachtae* Sateesh, Bhat, and Devaki are fungus of the *Phomopsis* genus that has been identified as the fungus responsible for the dieback in *Azadirachta indica* (Neem) in the regions of India. Neem is a very hardy and drought tolerant tree, has antifungal and antibacterial properties, but it still has many diseases. Different pathogens affect every part of the tree: the twigs, shoots, leaves, roots, and bark. Scientists have isolated both the fungus and the toxin. They have identified the pathogen in the tissue of all affected trees. Proper knowledge of toxin chemistry and its role in pathogenesis requires further investigations. Some experts consider dieback as a group of diseases with incompletely understood origins influenced by factors which predispose trees under stress to invasion.

Top dying disease is a disease of *Heritiera fomes*, the dominant species of mangrove tree growing in the Sundarbans. This disease has become more prevalent since about 1970 and may be linked to an increase in the heavy metal concentration of the sediment deposited in the Ganges Delta.

Hypothesis

The components of a forest ecosystem are complex and identifying specific cause-effect relationships between dieback and the environment is difficult. Etiology is the science of identifying the causes of death. Because there is no one single and clear cause of dieback there are multiple hypothesis for the causes and effects of dieback. The following hypotheses were agreed upon between the scientific exchanges of Germany and the United States in 1988.

Soil-Acidification/Aluminum-Toxicity: As soil becomes more acidic there is a release of aluminum that damages the tree's roots. Observed effects are: reduction of uptake and transport of some cations, reduction in root respiration, damage to fine feeder roots and root morphology, and reduction in elasticity of the cell walls. This is proposed by Professor Bernhard Ulrich in 1979.

Complex High-Elevation Disease: The combination of high ozone levels, acid deposition and nutrient deficiencies at high elevations kill trees. High ozone concentrations damage the leaves and needles of trees. Nutrients are leached from the foliage. The chain of events is magnified over time. This is proposed by a group of professors: Bernhard Prinz, Karl Rehfuess, Heinz Zöttle.

Red-Needle Disease of Spruce: The symptoms are needle drop and crown thinning. Needles turn a rust color and fall off. This is caused by foliar fungi that are secondary parasites attacking weakened trees. This is proposed by Professor Karl Rehfuess.

General Stress: The increased level of atmospheric pollutants hurts the root system and accumulates toxins in new leaves. Pollutants alter the growth, reduce the photoshythetic activity, and reduce the formation of secondary metabolites. It is believed that low concentrations levels are toxic. This is proposed by a group of professors led by Peter Schütt.

Excess Nitrogen Deposition: The increased level of nitrogen and ammonium, a fertilizer, could have the following possible effects: it could inhibit beneficial fungi, delay chemical reactions, disurbe normal balances between shoot growth and root growth, increase soil leaching. There is no experimental proof. This is proposed by Carl Olaf Tamm.

Organic Air Pollutants: This is similar to the general stress hypothesis, but focuses on organic compounds. The three compounds seriously discussed are ethylene, aniline, and dinitrophenol. Even at low levels these organic chemical compounds have caused: abnormal dropping of foliage, twisted foliage, and killing of seedlings. This is proposed by Fritz Führ.

Global Climate Change

Changes in mean annual temperature and drought are major contributing factors to forest dieback. As more carbon is released from dead trees, especially in the Amazon and Boreal forests, more greenhouse gases are released into the atmosphere. Increased levels of greenhouse gases increase the temperature of the atmosphere. The feedback loop is reinforced and the biological adaptations of the species determine its survival. Projections for dieback vary, but the threat of global climate change only stands to increase the rate of dieback.

Scientists do not know the exact tipping points of climate change and can only estimate the timescales. When a tipping point- the critical threshold- is reached, a small change in human activity can have long-term consequences on the environment. Two of the nine tipping points for major climate changes forecast for the next century are directly related to forest diebacks. Scientists are worried that forest dieback in the Amazon rain forest and the Boreal evergreen forest will trigger a tipping point in the next 50 years.

References

- Rudel, T.K. 2005 "Tropical Forests: Regional Paths of Destruction and Regeneration in the Late 20th Century" Columbia University Press ISBN 0-231-13195-X
- Leakey, Richard and Roger Lewin, 1996, The Sixth Extinction : Patterns of Life and the Future of Humankind, Anchor, ISBN 0-385-46809-1
- Taylor, Leslie (2004). The Healing Power of Rainforest Herbs: A Guide to Understanding and Using Herbal Medicinals. Square One. ISBN 9780757001444.
- Norman F. Cantor (9 June 1994). The civilization of the Middle Ages: a completely revised and expanded edition of Medieval history, the life and death of a civilization. HarperCollins. p. 564. ISBN 978-0-06-092553-6. Retrieved 4 December 2011.
- Ron Nielsen, The Little Green Handbook: Seven Trends Shaping the Future of Our Planet, Picador, New York (2006) ISBN 978-0-312-42581-4
- John F. Mongillo; Linda Zierdt-Warshaw (2000). Linda Zierdt-Warshaw, ed. Encyclopedia of environmental science. University of Rochester Press. p. 104. ISBN 978-1-57356-147-1.
- Stenstrup, Allen (2010). Forests. Greensboro, North Carolina: Morgan Reynolds Publishing. p. 89. ISBN 978-1-59935-116-2.
- M. Patricia Marchak (18 September 1995). Logging the globe. McGill-Queen's Press – MQUP. pp. 157–. ISBN 978-0-7735-1346-4. Retrieved 4 December 2011.

6

Mitigation Measures for Deforestation

Taking measure for deforestation is very necessary, considering the issues deforestation can cause. Forest restoration and afforestation are the main ways of measurement. While forest restoration is the recovery of forest structures, afforestation is the installation of plants in an area that has previously had no trees. This section serves as a source in understanding the measures taken for deforestation.

Forest Restoration

In the 1980s, conservation organizations warned that, once destroyed, tropical forests could never be restored. Thirty years of restoration research now challenge this: a) This site in Doi Suthep-Pui National Park, N. Thailand was deforested, over-cultivated and then burnt. The black tree stump was one of the original forest trees. Local people teamed up with scientists to repair their watershed.

Forest restoration is defined as "actions to re-instate ecological processes, which accelerate recovery of forest structure, ecological functioning and biodiversity levels towards those typical of climax forest" i.e. the end-stage of natural forest succession. Climax forests are relatively stable ecosystems that have developed the maximum biomass, structural complexity and species diversity that are possible *within the limits imposed by climate and soil and without continued disturbance from humans* (more explanation here). Climax forest is therefore the target ecosystem, which defines the ultimate aim of forest restoration. Since climate is a major factor that determines climax forest composition, global climate change may result in changing restoration aims.

Forest restoration is a specialized form of reforestation, but it differs from conventional tree plantations in that its primary goals are biodiversity recovery and environmental protection.

b) Fire prevention, nurturing natural regeneration and planting framework tree species resulted in trees growing above the weed canopy within a year.

c) After 12 years, the restored forest overwhelmed the black tree stump.

Scope

Forest restoration may include simply protecting remnant vegetation (fire prevention, cattle exclusion etc.) or more active interventions to accelerate natural regeneration, as well as tree planting and/or sowing seeds (direct seeding) of species characteristic of the target ecosystem. Tree species planted (or encouraged to establish) are those that are typical of, or provide a critical ecological function in, the target ecosystem. However, wherever people live in or near restoration sites, restoration projects often include economic species amongst the planted trees, to yield subsistence or cash-generating products.

Forest restoration is an inclusive process, which depends on collaboration among a wide range of stakeholders including local communities, government officials, non-government organizations, scientists and funding agencies. Its ecological success is measured in terms of increased biological diversity, biomass, primary productivity, soil organic matter and water-holding capacity, as well as the return of rare and keystone species, characteristic of the target ecosystem. Economic indices of success include the value of forest products and ecological services generated (e.g. watershed protection, carbon storage etc.), which ultimately contribute towards poverty reduction. Payments for such ecological services (PES) and forest products can provide strong incentives for local people to implement restoration projects.

Opportunities for Forest Restoration

Demonstration forest restoration plot, SUNY-ESF, Syracuse, NY

Forest restoration is appropriate wherever biodiversity recovery is one of the main goals of reforestation, such as for wildlife conservation, environmental protection, eco-tourism or to supply a wide variety of forest products to local communities. Forests can be restored in a wide range of circumstances, but degraded sites within protected areas are a high priority, especially where some climax forest remains as a seed source within the landscape. Even in protected areas, there are often large deforested sites: logged over areas or sites formerly cleared for agriculture. If protected areas are to act as Earth's last wildlife refuges, restoration of such areas will be needed.

Many restoration projects are now being implemented under the umbrella of "forest landscape restoration" (FLR), defined as a "planned process to regain ecological integrity and enhance human well-being in deforested or degraded landscapes". FLR recognizes that forest restoration has social and economic functions. It aims to achieve the best possible compromise between meeting both conservation goals and the needs of rural communities. As human pressure on landscapes increases, forest restoration will most commonly be practiced within a mosaic of other forms of forest management, to meet the economic needs of local people.

Natural Regeneration

Tree planting is not always essential to restore forest ecosystems. A lot can be achieved by studying how forests regenerate naturally, identifying the factors that limit regeneration and devising methods to overcome them. These can include weeding and adding fertilizer around natural tree seedlings, preventing fire, removing cattle and so on. This is "accelerated" or "assisted" natural regeneration. It is simple and cost-effective, but it can only operate on trees that are already present, mostly light-loving pioneer species. Such tree species are not usually those that comprise climax forests, but they can foster recolonization of the site by shade-tolerant climax forest tree species, via natural seed dispersal from remnant forest. Because this is a slow process, biodiversity recovery can usually be accelerated by planting some climax forest tree species, especially large-seeded, poorly dispersed species. It is not feasible to plant all the tree species that may have formerly grown in the original primary forest and it is usually unnecessary to do so, if the framework species method can be used.

Post-fire Regeneration

In large parts of the world, forest fires cover a heavy toll on forests. That can be because of provoked deforestation in order to substitute forests by crop areas, or in dry areas, because of wild fires occurring naturally or intentionally. A whole section of forest landscape restoration in linked to this particular problem, as in many cases, the net loss of ecosystem value is very high and can open the drop to an accelerated further degradation of the soil conditions through erosion and desertification. This indeed has dire consequences on both the quality of the habitats and their related fauna. Nevertheless, in some specific cases, wild fires do actually allow to increase the biodiversity index of the burnt area, in which case the Forest Restoration Strategies tend to look for a different land-use.

Forest Restoration Projects

Ashland Forest Resiliency Stewardship Project

The Ashland Forest Resiliency Stewardship Project (AFR) is a decade long, science-based project launched in 2010 with the intent of reducing severe wildfire risk, but also protecting water quality, old-growth forest, wildlife, people, property, and the overall quality of life within the Ashland watershed. The primary stakeholders in this cooperative restoration effort are the U.S. Forest Service, the City of Ashland, Lomaktsi Restoration Project, and the Nature Conservancy. The project was launched with initial funding from the Economic Recovery stimulus, and has more recently received funding from the Forest Service Hazardous Fuels program and the Joint Chiefs Landscape Restoration Partnerships program to back the project through 2016 .

Located in the dry forests of southern Oregon, the threat of wildfire is a reality for land managers and property owners alike. The boundaries of the city of Ashland intersect with the surrounding forest in what is referred to as the wildland-urban interface (WUI). Historically, the forests of this region experienced a relatively frequent fire return interval, which prevented buildup of heavy fuel loads. A century of fire exclusion and suppression on federal lands in the Pacific Northwest has led to increased forest density and fuel loads, and thus a more persistent threat of devastating wildfire.

The AFR project has implemented restoration techniques and prescriptions that aim to replicate the process of ecological succession in dry, mixed-conifer forests of the Pacific

Northwest. The approach involves a combination of fuels reduction, thinning small-diameter trees, and carrying out prescribed burns. Priority is given to maintaining ecological function and complexity by retaining the largest and oldest trees, preserving wildlife habitat and riparian areas, and protecting erodible soils and maintaining slope stability.

Since its inception in 2010, the AFR project has completed restoration work on 4,000 of the 7,600 acres slated for the project. The project has provided educational experience to over 2,000 students and has benefitted the local community by creating jobs and providing workforce training. Currently, helicopter logging operations are thinning 1,100 acres of the watershed while controlled burning operations take place as air quality conditions allow.

Forest Landscape Restoration

Forest landscape restoration (FLR) is defined as "a planned process to regain ecological integrity and enhance human well-being in deforested or degraded landscapes". It comprises tools and procedures to integrate site-level forest restoration actions with desirable landscape-level objectives, which are decided upon via various participatory mechanisms among stakeholders. The concept has grown out of collaboration among some of the world's major international conservation organizations including the International Union for Conservation of Nature (IUCN), the World Wide Fund for Nature (WWF), the World Resources Institute and the International Tropical Timber Organization (ITTO).

Aims

The concept of FLR was conceived to bring about compromises between meeting the needs of both humans and wildlife, by restoring a range of forest functions at the landscape level. It includes actions to strengthen the resilience and ecological integrity of landscapes and thereby keep future management options open. The participation of local communities is central to the concept, because they play a critical role in shaping the landscape and gain significant benefits from restored forest resources. Therefore, FLR activities are inclusive and participatory.

Desirable Outcomes

The desirable outcomes of an FLR program usually comprise a combination of the following, depending on local needs and aspirations:

- identification of the root causes of forest degradation and prevention of further deforestation,
- positive engagement of people in the planning of forest restoration, resolution of land-use conflicts and agreement on benefit-sharing systems,

- compromises over land-use trade-offs that are acceptable to the majority of stakeholders,
- a repository of biological diversity of both local and global value,
- delivery of a range of utilitarian benefits to local communities including:
    - a reliable supply of clean water,
    - environmental protection particularly watershed services (e.g. reduced soil erosion, lower landslide risk, flood/drought mitigation etc.),
    - a sustainable supply of a diverse range of forest products including foods, medicines, firewood etc.,
    - monetary income from various sources e.g. ecotourism, carbon trading via the REDD+ mechanism and from payments for other environmental services (PES)

Activities

FLR combines several existing principles and techniques of development, conservation and natural resource management, such as landscape character assessment, participatory rural appraisal, adaptive management etc. within a clear and consistent evaluation and learning framework. An FLR program may comprise various forestry practices on different sites within the landscape, depending on local environmental and socioeconomic factors. These may include protection and management of secondary and degraded primary forests, standard forest restoration techniques such as “assisted” or “accelerated” natural regeneration (ANR) and the planting of framework tree species to restore degraded areas, as well as conventional tree plantations and agroforestry systems to meet more immediate monetary needs

The IUCN hosts the The Global Partnership on Forest Landscape Restoration, which co-ordinates development of the concept around the world.

In 2014, the Food and Agricultural Organization of the United Nations established the Forest and Landscape Restoration Mechanism. The Mechanism supports countries to implement FLR as a contribution to achieving the Bonn Challenge - the restoration of 150 million hectare of deforested and degraded lands by 2020 - and the Convention on Biological Diversity Aichi Biodiversity Targets - related to ecosystem conservation and restoration.

In partnership with the Global Mechanism of the United Nations Convention to Combat Desertification, FAO released two discussion papers on sustainable financing for FLR in 2015. *Sustainable Financing for Forest and Landscape Restoration: The Role of Public Policy Makers* provides recommendations and examples of FLR financing for countrues. *Sustainable Financing for Forest and Landscape Restoration - Opportu-*

nities, challenges and the way forward provides an overview of funding sources and financial instruments available for FLR activities.

Reforestation

Tropical tree nursery at Planeta Verde Reforestación S.A.'s plantation in Vichada Department, Colombia

A 15-year-old reforested plot of land

A 21-year-old plantation of red pine in Southern Ontario

Reforestation is the natural or intentional restocking of existing forests and woodlands that have been depleted, usually through deforestation. Reforestation can be used to rectify or improve the quality of human life by soaking up pollution and dust from the air, rebuild natural habitats and ecosystems, mitigate global warming since forests facilitate biosequestration of atmospheric carbon dioxide, and harvest for resources, particularly timber, but also non-timber forest products.

The term reforestation is similar to afforestation, the process of restoring and recreating areas of woodlands or forests that may have existed long ago but were deforested or otherwise removed at some point in the past. Sometimes the term *re-afforestation* is used to distinguish between the original forest cover and the later re-growth of forest to an area. Forestation is the establishment of forest growth on areas that either had forest or lacked it.

Special tools, e.g. tree planting bars, are used to make planting of trees easier and faster.

Management

A debated issue in managed reforestation is whether or not the succeeding forest will have the same biodiversity as the original forest. If the forest is replaced with only one species of tree and all other vegetation is prevented from growing back, a monoculture forest similar to agricultural crops would be the result. However, most reforestation involves the planting of different seedlots of seedlings taken from the area, often of multiple species. Another important factor is the natural regeneration of a wide variety of plant and animal species that can occur on a clear cut. In some areas the suppression of forest fires for hundreds of years has resulted in large single aged and single species forest stands. The logging of small clear cuts and/or prescribed burning, actually increases the biodiversity in these areas by creating a greater variety of tree stand ages and species.

For Harvesting

Reforestation need not be only used for recovery of accidentally destroyed forests. In some countries, such as Finland, many of the forests are *managed* by the wood products and pulp and paper industry. In such an arrangement, like other crops, trees are planted to replace those that have been cut. In such circumstances, the industry can cut the trees in a way to allow easier reforestation. The wood products industry systematically replaces many of the trees it cuts, employing large numbers of summer workers for tree planting work. For example, in 2010, Weyerhaeuser reported planting 50 million seedlings. However replanting an old-growth forest with a plantation is not replacing the old with the same characteristics in the new.

In just 20 years, a teak plantation in Costa Rica can produce up to about 400 m^3 of wood per hectare. As the natural teak forests of Asia become more scarce or difficult to obtain, the prices commanded by plantation-grown teak grows higher every year. Other species such as mahogany grow more slowly than teak in Tropical America but are also extremely valuable. Faster growers include pine, eucalyptus, and *Gmelina*.

Reforestation, if several indigenous species are used, can provide other benefits in addition to financial returns, including restoration of the soil, rejuvenation of local flora and fauna, and the capturing and sequestering of 38 tons of carbon dioxide per hectare per year.

The reestablishment of forests is not just simple tree planting. Forests are made up of a community of species and they build dead organic matter into soils over time. A major tree-planting program could enhance the local climate and reduce the demands of burning large amounts of fossil fuels for cooling in the summer.

For Climate Change Mitigation

Forests are an important part of the global carbon cycle because trees and plants absorb carbon dioxide through photosynthesis. By removing this greenhouse gas from the air, forests function as terrestrial carbon sinks, meaning they store large amounts of carbon. At any time, forests account for as much as double the amount of carbon in the atmosphere. Even as more anthropogenic carbon is produced, forests remove around three billion tons of anthropogenic carbon every year. This amounts to about 30% of all carbon dioxide emissions from fossil fuels. Therefore, an increase in the overall forest cover around the world would tend to mitigate global warming.

There are four major strategies available to mitigate carbon emissions through forestry activities: increase the amount of forested land through a reforestation process; increase the carbon density of existing forests at a stand and landscape scale; expand the use of forest products that will sustainably replace fossil-fuel emissions; and reduce carbon emissions that are caused from deforestation and degradation.

Achieving the first strategy would require enormous and wide-ranging efforts. However, there are many organizations around the world that encourage tree-planting as a way to offset carbon emissions for the express purpose of fighting climate change. For example, in China, the Jane Goodall Institute, through their Shanghai Roots & Shoots division, launched the Million Tree Project in Kulun Qi, Inner Mongolia to plant one million trees to stop desertification and help curb climate change. China has used 24 billion metres squared of new forest plantation and natural forest regrowth to offset 21% of Chinese fossil fuel emissions in 2000. In Java, Indonesia each newlywed couple is to give whoever is sermonizing their wedding 5 seedlings to combat global warming. Each couple that wishes to have a divorce has to give 25 seedlings to whoever divorces them.

The second strategy has to do with selecting species for tree-planting. In theory, planting any kind of tree to produce more forest cover would absorb more carbon dioxide from the atmosphere. On the other hand, a genetically modified tree specimen might grow much faster than any other regular tree. Some of these trees are already being developed in the lumber and biofuel industries. These fast-growing trees would not only be planted for those industries but they can also be planted to help absorb carbon dioxide faster than slow-growing trees.

Extensive forest resources placed anywhere in the world will not always have the same impact. For example, large reforestation programs in boreal or subarctic regions have a limited impact on climate mitigation. This is because it substitutes a bright snow-dominated region that reflects the sunlight with dark forest canopies. A study from the National Center for Atmospheric Research in Boulder, Colorado, USA, found that trees in temperate latitudes have a net warming effect on the atmosphere. The heat that dark leaves release without absorbing outweighs the carbon they sequester. On the other hand, a positive example would be reforestation projects in tropical regions, which would lead to a positive biophysical change such as the formation of clouds. These clouds would then reflect the sunlight, creating a positive impact on climate mitigation.

There is an advantage to planting trees in tropical climates with wet seasons. In such a setting, trees have a quicker growth rate because they can grow year-round. Trees in tropical climates have, on average, larger, brighter, and more abundant leaves than non-tropical climates. A study of the girth of 70,000 trees across Africa has shown that tropical forests are soaking up more carbon dioxide pollution than previously realized. The research suggests almost one fifth of fossil fuel emissions are absorbed by forests across Africa, Amazonia and Asia. Simon Lewis, a climate expert at the University of Leeds, who led the study, said: "Tropical forest trees are absorbing about 18% of the carbon dioxide added to the atmosphere each year from burning fossil fuels, substantially buffering the rate of change."

It is also important to deal with the rate of deforestation. At this point, there are 13 billion metres squared of tropical regions that are deforested every year. There is potential for these regions to reduce rates of deforestation by 50% by 2050, which would be a huge contribution to stabilize the global climate.

On Abandoned Farmland

With increased intensive agriculture and urbanization, there is an increase in the amount of abandoned farmland. By some estimates, for every half a hectare of original old-growth forest cut down, more than 20 hectares of new secondary forests are growing, even though they do not have the same biodiversity as the original forests and original forests store 60% more carbon than these new secondary forests. According to a study in Science, promoting regrowth on abandoned farmland could offset years of carbon emissions.

Promotion Strategies

Land Rights

Transferring land rights to indigenous inhabitants is argued to efficiently conserve forests.

Transferring rights over land from public domain to its indigenous inhabitants is argued to be a cost effective strategy to conserve forests. This includes the protection of such rights entitled in existing laws, such as India's Forest Rights Act. The transferring of such rights in China, perhaps the largest land reform in modern times, has been argued to have increased forest cover. In Brazil, forested areas given tenure to indigenous groups have even lower rates of clearing than national parks.

Incentives

Some incentives for reforestation can be as simple as a financial compensation. Streck and Scholz (2006) explain how a group of scientists from various institutions have developed a compensated reduction of deforestation approach which would reward developing countries that disrupt any further act of deforestation. Countries that participate and take the option to reduce their emissions from deforestation during a committed period of time would receive financial compensation for the carbon dioxide emissions that they avoided. To raise the payments, the host country would issue government bonds or negotiate some kind of loan with a financial institution that would want to take part in the compensation promised to the other country. The funds received by the country could be invested to help find alternatives to the extensive cutdown of forests. This whole process of cutting emissions would be voluntary, but once the country has agreed to lower their emissions they would be obligated to reduce their emissions. However, if a country was not able to meet their obligation, their target would get added to their next commitment period. The authors of these proposals see this as a solely government-to-government agreement; private entities would not participate in the compensation trades.

Examples

Forest regrowth in Mount Baker-Snoqualmie National Forest, Washington state, USA

Sub-Saharan Africa

Plans to plant a nine-mile width of trees on the Southern Border of the Sahara Desert. The Great Green Wall initiative is a pan-African proposal to "green" the continent from west to east in order to battle desertification. It aims at tackling poverty and the degradation of soils in the Sahel-Saharan region, focusing on a strip of land of 15 km (9 mi) wide and 7,500 km (4,750 mi) long from Dakar to Djibouti.

Canada

In Canada, overall forest cover is increasing over the last decades.

China

In China, extensive replanting programs have existed since the 1970s. Programs have had overall success. The forest cover has increased from 12% of China's land area to 16%. However, specific programs have had limited success. The "Green Wall of China", an attempt to limit the expansion of the Gobi Desert is planned to be 2,800 miles (4,500 km) long and to be completed in 2050. China plans to plant 26 billion trees in the next decade that is two trees for every Chinese citizen per year. China requires that students older than 11 years old plant one tree a year until their high school graduation.

Germany

Reforestation is required as part of the federal forest law. 31% of Germany is forested, according to the second forest inventory of 2001–2003. The size of the forest area in Germany increased between the first and the second forest inventory due to forestation of degenerated bogs and agricultural areas.

India

Jadav Payeng had received national awards for reforestation efforts, known as the "Molai forest".

Japan

Mass environmental and human-body pollution along with relating deforestation, water pollution, smoke damage, and loss of soils caused by mining operations in Ashio, Tochigi became the first environmental social issue in Japan, efforts by Shōzō Tanaka had grown to large campaigns against copper operation. This led to the creation of 'Watarase Yusuichi Pond', to settle the pollution which is a Ramsar site today. Reforestation was conducted as a part of afforestation due to inabilities of self-recovering by the natural land itself due to serious soil pollution and loss of woods consequence in loss of soils for plants to grow, thus needing artificial efforts involving introducing of healthy soils from outside. Starting since in 1897, about 50% of once bald mountains backed to green.

Lebanon

For thousands of years Lebanon was covered by forests, one particular species of interest, *Cedrus libani* or Lebanon Cedar was exceptionally valuable timber species. Virtually every ancient culture that shared the Mediterranean Sea harvested these trees, the Ancient Egyptians, the Greeks, the Roman Empire, early Christians, the Muslims, Persians, Assyrians and Babylonians. Extensive deforestation has occurred, with only small remnants of the original forests surviving. Deforestation has been particularly severe in Lebanon and on Cyprus. The Lebanon Reforestation Initiative aims to restore Lebanon's native forests. Projects financed locally and by international charity are performing extensive reforestation of cedar being carried out in the Mediterranean region, particularly in Lebanon and Turkey, where over 50 million young cedars are being planted annually.

Philippines

The Philippines increased its forest area to 240,000 hectares (593,053 acre) per year from 2010 to 2015. In 2011-2015 period, reforestation through the government's National Greening Program covered 1,258,692 hectares of land, which is 105 percent more than the 1,200,000 hectares planting target for the same period. An estimated eight million hectares of land nationwide need reforestation and rehabilitation, according to the Forest Management Bureau.

United States

It is the stated goal of the US Forest Service to manage forest resources sustainably. This includes reforestation after timber harvest, among other programs.

Organizations

Trees for the Future has assisted more than 170,000 families, in 6,800 villages of Asia, Africa and the Americas, to plant over 35 million trees.

Wangari Maathai, 2004 Nobel Peace Prize recipient, founded the Green Belt Movement which planted over 47 million trees to restore the Kenyan environment.

Shanghai Roots & Shoots, a division of the Jane Goodall Institute, launched The Million Tree Project in Kulun Qi, Inner Mongolia to plant one million trees to stop desertification and alleviate global warming.

Criticisms

Reforestation competes with other land uses such as food production, livestock grazing and living space for further economic growth. Reforestation often has the tendency to create large fuel loads, resulting in significantly hotter combustion than fires involving low brush or grasses. Reforestation can divert large amounts of water from other activities. Reforesting sometimes results in extensive canopy creation that prevents growth of diverse vegetation in the shadowed areas and generating soil conditions that hamper other types of vegetation. Trees used in some reforesting efforts (e.g., *Eucalyptus globulus*) tend to extract large amounts of moisture from the soil, preventing the growth of other plants.

There is also the risk that through a forest fire or insect outbreak much of the stored carbon in a reforested area could make its way back to the atmosphere. Reduced harvesting rates and fire suppression have caused an increase in the forest biomass in the western United States over the past century. This causes an increase of about a factor of four in the frequency of fires due to longer and hotter dry seasons.

Secondary Forest

The forest in Stanley Park, Vancouver, Canada is generally considered to have second and third growth characteristics. This photo shows regeneration, a tree growing out of the stump of another tree that was felled in 1962 by the remnants of Typhoon Freda.

A secondary forest (or second-growth forest) is a forest or woodland area which has re-grown after a major disturbance such as fire, insect infestation, timber harvest or windthrow, until a long enough period has passed so that the effects of the disturbance are no longer evident. It is distinguished from an old-growth forest (primary or primeval forest), which has not undergone such disruptions, as well as third-growth forests that result from severe disruptions in second growth forests.

Description

Depending on the forest, the development of primary characteristics may take anywhere from a century to several millennia. Hardwood forests of the eastern United States, for example, can develop primary characteristics in one or two generations of trees, or 150–500 years. Often the disruption is the result of human activity, such as logging, but natural phenomena that produce the same effect are often included in the definition. Secondary forests tend to have trees closer spaced than primary forests and contain less undergrowth than primary forests. Secondary forests typically were thought to lack biodiversity compared to primary forests, however this has been challenged in recent years. Usually, secondary forests have only one canopy layer, whereas primary forests have several.

Secondary forestation is common in areas where forests have been lost by the slash-and-burn method, a component of some shifting cultivation systems of agriculture. Secondary forests may also arise from forest that has been harvested heavily or over a long period of time, forest that is naturally regenerating from fire and from abandoned pastures or areas of agriculture. It takes a secondary forest typically forty to 100 years to begin to resemble the original old-growth forest; however, in some cases a secondary forest will not succeed, due to erosion or soil nutrient loss in certain tropical forests.

Secondary forests re-establish by the process of succession. Openings created in the forest canopy allow sunlight to reach the forest floor. An area that has been cleared will first be colonized by pioneer species. Even though some species loss may occur with primary forest removal, a secondary forest can protect the watershed from further erosion and provides habitat. Secondary forests may also buffer edge effects around mature forest fragments and increase connectivity between them. They may also be a source of wood and other forest products.

Today most of the forest of the United States, the eastern part of North America and Europe consist of secondary forest.

Rainforests

In the case of tropical rainforests, where soil nutrient levels are characteristically low, the soil quality may be significantly diminished following the removal of primary forest. In Panama, growth of new forests from abandoned farmland exceeded loss of primary rainforest in 1990. However, due to the diminished quality of soil, among other factors, the presence of a significant majority of primary forest species fail to recover in these second-growth forests.

Afforestation

An afforestation project in Rand Wood, Lincolnshire, England

Afforestation is the establishment of a forest or stand of trees in an area where there was no previous tree cover. Reforestation is the reestablishment of forest cover, either naturally (by natural seeding, coppice, or root suckers) or artificially (by direct seeding or planting).

Forestation is the establishment of forest growth on areas that either had forest or lacked it. Reforestation and afforestation are categories of forestation. Many governments and non-governmental organizations directly engage in programs of *afforestation* to create forests, increase carbon capture and carbon sequestration, and help to anthropogenically improve biodiversity. (In the UK, afforestation may mean converting the legal status of some land to "royal forest".) Special tools, e.g. tree planting bar, are used to make planting of trees easier and faster.

Biological Process

Gap dynamics is the pattern of plant growth that occurs following the creation of a forest gap, a local area of natural disturbance that results in an opening in the canopy of a forest. Gap dynamics are a typical characteristic of temperate and tropical forests, and have a wide variety of causes and effects on forest life.

In Areas of Degraded Soil

In some places, forests need help to reestablish themselves because of environmental factors. For example, in arid zones, once forest cover is destroyed, the land may dry and become inhospitable to new tree growth. Other factors include overgrazing by livestock, especially animals such as goats, cows, and over-harvesting of forest resources. Together these may lead to desertification and the loss of topsoil; without soil, forests cannot grow until the long process of soil creation has been completed - if erosion allows this. In some tropical areas, forest cover removal may result in a duricrust or duripan that effectively seal off the soil to water penetration and root growth. In many areas, reforestation is impossible because people are using the land. In other areas, mechanical breaking up of duripans or duricrusts is necessary, careful and continued watering may be essential, and special protection, such as fencing, may be needed.

Countries and Regions

Afforested botanical garden in Hattori Ryokuchi Park, Japan.

Brazil

There is extensive and ongoing Amazon deforestation.

China

China has deforested most of its historically wooded areas. China reached the point where timber yields declined far below historic levels, due to over-harvesting of trees beyond sustainable yield. Although it has set official goals for reforestation, these goals are set over an 80-year time horizon and have not been significantly met by 2008. China is trying to correct these problems by projects as the Green Wall of China, which aims to replant a great deal of forests and halt the expansion of the Gobi desert. A law promulgated in 1981 requires that every school student over the age of 11 plants at least one tree per year. As a result, China has the highest afforestation rate of any country or region in the world, with 47,000 square kilometers of afforestation in 2008. However, the forest area per capita is still far lower than the international average. There has also been considerable criticism regarding the effectiveness of planting so many trees especially in regions where they never grew prior. Studies reveal that the water table of those areas is becoming deeper indicating significant water loss.

India

Afforestation in South India

India has witnessed a minor increase in the percentage of the land area under forest cover from 1950 to 2006. In 1950 around 40.48 million hectares was covered by forest. In 1980 it increased to 67.47 million hectares and in 2006 it was found to be 69 million hectares. 23% of India is covered by forest. The forests of India are grouped into 5 major categories and 16 types based on biophysical criteria. 38% of forest is categorised as subtropical dry deciduous and 30% as tropical moist deciduous plus other smaller groups. It is taken care that only local species are planted in an area. Trees bearing fruits are preferred wherever possible due to their function as a food source.

Hong Kong

Since the founding of the crown colony in the 19th century, afforestation has taken place to prevent soil erosion in the catchment areas of the reservoirs that were built. During the Japanese occupation in the Second World War, the countryside was deforested as the remaining population required fuel to survive. Most of the trees were cut down and extensive reafforestation was carried out after the war. Trees that were planted are mostly non-native species, such as: Pinus massoniana, Acacia confusa (Formosan acacia), Lophostemon confertus and the Paper Bark Tree.

Burkina Faso

Desertification is increasing along the Sahel, the strip of land between Africa's fertile tropics and the Sahara Desert. After a crippling famine in the 1970s caused by overgrazing and deforestation, a local community approach has been pioneered by Yacouba Sawadogo, a peasant farmer. By replanting trees and crops together in holes filled with compost, whole villages have been able to move back to areas considered uninhabitable.

Iran

Iran is considered a low forest cover region of the world with present cover approximating seven percent of the land area. This is a value reduced by an estimated six million hectares of virgin forest, which includes oak, almond and pistachio. Due to soil substrates, it is difficult to achieve afforestation on a large scale compared to other temperate areas endowed with more fertile and less rocky and arid soil conditions. Consequently, most of the afforestation is conducted with non-native species, leading to habitat destruction for native flora and fauna, and resulting in an accelerated loss of biodiversity.

JNF trees in the Negev Desert. Man-made dunes (here a liman) help keep in rainwater, creating an oasis.

Israel

Tree-planting is an ancient Jewish tradition, mentioned in the Talmud as being more important than greeting the Messiah. With over 240 million planted trees, Israel is one of only two countries that entered the 21st century with a net gain in the number of trees, due to massive afforestation efforts. Israeli forests are the product of a major afforestation campaign by the Jewish National Fund (JNF).

Critics argue that many JNF lands inside the West Bank were illegally confiscated from Palestinian refugees, and that the JNF furthermore should not be involved with lands in the West Bank. Shaul Ephraim Cohen has claimed that trees have been planted to restrict Bedouin herding. Susan Nathan wrote that forests were planted on the site of abandoned Arab villages after the 1948 war.

Since 2009, the JNF has provided the Palestinian Authority with 3,000 tree seedlings for a forested area being developed on the edge of the new city of Rawabi, north of Ramallah.

North Africa

In North Africa, the Sahara Forest Project coupled with the Seawater greenhouse has been proposed. Some projects have also been launched in countries as Senegal to revert desertification. As of 2010, African leaders are discussing the combining of national resources to increase effectiveness. In addition, other projects as the Keita Project in Niger have been launched in the past, and have been able to locally revert damage done by desertification.

Europe

Europe has deforested the majority of its historical forests. The European Union (EU) has paid farmers for afforestation since 1990, offering grants to turn farmland back into forest and payments for the management of forest. Between 1993 and 1997, EU afforestation policies made possible the re-forestation of over 5,000 square kilometres of land. A second program, running between 2000 and 2006, afforested more than 1000 square kilometres of land (precise statistics not yet available). A third such program began in 2007. Europe's forests are growing by 0.8 million ha a year thanks to these programmes.

In Poland, the National Program of Afforestation was introduced by the government after World War II, when area of forests shrank to 20% of country's territory. Consequently, forested areas of Poland grew year by year, and on December 31, 2006, forests covered 29% of the country. It is planned that by 2050, forests will cover 33% of Poland.

According to Food and Agriculture Organization statistics, Spain had the third fastest afforestation rate in Europe in the 1990-2005 period, after Iceland and Ireland. In those years, a total of 44,360 square kilometers were afforested, and the total forest cover rose from 13,5 to 17,9 million hectares. In 1990, forests covered 26.6% of the Spanish territory. As of 2007, that figure had risen to 36.6%. Spain today has the fifth largest forest area in the European Union.

In January 2013 the UK government set a target of 12% woodland cover in England by 2060, up from the then 10%. Government-backed initiatives such as the Woodland Carbon Code are intended to support this objective by encouraging corporations and landowners to create new woodland to offset their carbon emissions.

Alpine and Subalpine regions have undergone a lot of deforestation and then forestation in the last 300 years. Out of this has emerged much practical experience. One example is the Rotten group, which is a method to bring in stable age mixed tree communities.

Australia

In Adelaide, South Australia (a city of 1.3 million), Premier Mike Rann (2002 to 2011) launched an urban forest initiative in 2003 to plant 3 million native trees and shrubs by 2014 on 300 project sites across the metro area. The projects range from large habitat restoration projects to local biodiversity projects. Thousands of Adelaide citizens have participated in community planting days. Sites include parks, reserves, transport corridors, schools, water courses and coastline. Only trees native to the local area are planted to ensure genetic integrity. Premier Rann said the project aimed to beautify and cool the city and make it more liveable; improve air and water quality and reduce Adelaide's greenhouse gas emissions by 600,000 tonnes of Co2 a year. He said it was also about creating and conserving habitat for wildlife and preventing species loss.

United States

The United States is roughly one-third covered in forest and woodland. Nevertheless, areas in the US were subject to significant tree planting. In the 1800s people moving westward encountered the Great Plains; land with fertile soil, a growing population and a demand for timber but with few trees to supply it. So tree planting was encouraged along homesteads. Arbor Day was founded in 1872 by Julius Sterling Morton in Nebraska City, Nebraska. By the 1930s the environmental disaster, the Dust Bowl signified a reason for significant tree cover. Public work's programs under the New Deal saw the planting of 18,000 miles of windbreaks stretching from North Dakota to Texas to fight soil erosion.

At their summit in Copenhagen in 2009, organised by the UK based The Climate Group, leaders of sub-national governments - States, Regions and Provinces - unanimously supported a recommendation by Premier Rann to plant 1 billion trees across their varied jurisdictions. The initiative was strongly supported by leaders present including Quebec Premier Jean Charest, California Governor Arnold Schwarzenegger and Scottish First Minister Alex Salmond. At a subsequent meeting in Rio de Janeiro in June 2012, The Climate Group announced that it had already received commitments by member governments to plant more than 500 million trees.

Forest Inventory

Forest inventory is the systematic collection of data and forest information for assessment or analysis. An estimate of the value and possible uses of timber is an important part of the broader information required to sustain ecosystems. When taking forest inventory the following are important things to measure and note: species, diameter at breast height (DBH), height, site quality, age, and defects. From the data collected one can calculate the number of trees per acre, the basal area, the volume of trees in an area, and the value of the timber. Inventories can be done for other reasons than just calculating the value. A forest can be cruised to visually assess timber and determine potential fire hazards and the risk of fire. The results of this type of inventory can be used in preventative actions and also awareness. Wildlife surveys can be undertaken in conjunction with timber inventory to determine the number and type of wildlife within a forest. The aim of the statistical forest inventory is to provide comprehensive information about the state and dynamics of forests for strategic and management planning. Merely looking at the forest for assessment is called taxation.

History

Surveying and taking inventory of trees originated in Europe in the late 18th century out of a fear that wood (the main source of fuel) would run out. The first information was organized into maps used to plan out usage. In the early 19th century forest harvesters estimated the volume and dispersal of trees within smaller forests with their eyes. More diverse and larger forests were divided into smaller sections of similar type trees that were individually estimated by visual inspection. These estimates were related together to figure out the entire forest's available resources. As the 19th century progressed so did the measurement techniques. New relationships between diameter, height, and volume were discovered and exploited. These newfound relationships allowed for a more accurate assessment of wood types and yields of much larger forests. By 1891, these surveys were conducted through sample-based methods involving statistical averages and more sophisticated measuring devices were implemented. In the 20th century, the statistical method of sampling had become well established and commonly used. Further developments, such as unequal probability sampling, arose. As the 20th century progressed, an understanding of co-efficients of error became clearer and the new technology of computers combined with the availability of aerial as well as satellite photography, further refined the process. Laser Scanning both terrestrially and aerially are now used alongside more manual methods. As a result, sampling accuracy and assessment values became more accurate and allowed for modern practices to arise.

A forest inventory does not only record the trees height, DBH and number for tree yield calculations. But it also records the conditions of the forest. So this would include, geology, site conditions, tree health and other forest factors.

Timber Cruise

A timber cruise is a sample measurement of a stand used to estimate the amount of standing timber that the forest contains. These measurements are collected at sample locations called plots, quadrants, or strips. Each of these individual sample areas is one observation in a series of observations called a sample. These sample areas are generally laid out in some random fashion usually in the form of a line plot survey. Depending on the size of the plot and the number of plots measured, the data gathered from these plots can then be manipulated to achieve varying levels of certainty for an estimate that can be applied to the entire timber stand. This estimate of stand conditions, species composition, volume and other measured attributes of a forest system can then be used for various purposes. For example, in British Columbia the sale of Crown timber is a business proposition and both the buyer and the Ministry of Forests and Range (seller) must know the quantity and the quality of timber being sold. Our satellite provides the essential data and information for determining stumpage rates, for establishing conditions of sale and for planning of the logging operations by the licensee. Generally a timber cruise includes measurements or estimates of timber volume by forest product sort (and sometimes grade), log defect, and log lengths, whether the estimates are made in the field or using computer software.

Stranded Examinations

Example of hardware equipment for forest inventories: GPS and laser rangefinder for mapping connected to a field rugged computer.

Before, the most common type of inventory is one that uses a random sampling technique which groups similar forests into one category based on age, stand structure, species, and location. The next step would be to begin from a random place and measure circular plots that are equidistant from one another. Inventory undertaken in conjunction with wildlife surveys has a regularized distance between plots. When taking inventory, a decision has to be made about which types of plots to measure. There are several different types of plots. The two most common types are *fixed radius* and *variable radius*, also known as *prism radius*. In a fixed radius plot, the forester finds the center of a plot and every tree within a certain fixed distance away from that point is measured. Variable radius plots are used more for inventory of volume. During this method, an angle is created and projected from the plot center and all trees that are larger than the projected right angle are measured with a ruler called a Tunagmetor .

Types of Sample Plots

Fixed Area Plot

Fixed area plot sample measurements are taken so that they are a fraction of the entire timber stand. This means that the numbers are all proportional to the actual stand values and that by multiplying by the correct corresponding value you can obtain the actual tract values. These plots are taken randomly so that each sample point has an equal probability of being included in the random sample. Commonly a 100m^2 plot is taken.

Variable Size Plots

A variable size plot is more dependent of the size of the trees. The tract is measured on a series of points and the trees are tallied for being in or out depending on their size and location relevant to the plot center. Usually an angle gauge, wedge prism or Relascope are used to gather data for this type of plot. This allows for a very quick estimate of the volume and species of a given tract.

Transects

Transects are arbitrarily determined lines (to prevent sampling bias) through a stand employed as a linear form of sample plot. They are sometimes referred to as "strip lines."

Plot Selection

Plots are samples of the forest being inventoried and so are selected according to what is looked for.

Simple Random Sampling

A computer or calculator random number generator is used to assign plots to be sampled. Here random means an equal chance of any plot being selected out of all of the plots available. It does not mean haphazard. Often it is modified to avoid sampling roads, ensure coverage of unsampled areas and for logistics of actually getting to the plots.

Systematic Sampling

Commonly this is done by a random point and then laying a grid over a map of the area to be sampled. This grid will have preassigned plot areas to be sampled. It means more efficient logistics and removes some of the human bias that may be there with simple random sampling.

Systematic Stratified Sampling

There is some HAPPY grouping, for example by age classes or soil characteristics or slope elevation. And then plots are chosen from each grouping by another sampling technique. It requires some knowledge of the land first and also trust that the groupings have been done properly. In forestry it may be done to separate plantation areas from mixed forest for example and reduce the amount of sampling time needed.

Systematic Clustered Sampling

When it is not possible to make strata for stratified sampling, there may be some knowledge about the forest where it can be said that small groupings are possible. These small groupings of plots if they are near to each other form a cluster. These clusters are then randomly sampled with the belief that they are representing the actual mix of the forest. As they are close to each other there is less waking needed and so it is more efficient.

Timber Metrics

The amount of standing timber that a forest contains is determined from:

- Age class. This is a misnomer as in German it is Wuchsklass (Growth class) and properly should be the size of the tree (Size class), which may be limited by shade over it and not necessarily the biological and thus physiological age of the tree. This distinction is important if tree growth over time is expected by an owner or forester to produce timber, as a small old tree will grow differently from a small young tree. Commonly these age classes are split into: Seedling, Sapling, Pole, Mature Tree (subdivided into Weak wood, Middle wood and Strong wood stages), Old / Scenescant Tree. Sometimes it is called size class or a cohort. There are differences between countries and forests.
- Basal area – defines the area of a given section of land that is occupied by the cross-section of tree trunks and stems at their base
- Diameter at breast height (DBH) – measurement of a tree's girth standardized with different countries having different standards they are often at 1.3 meters (about 4.5 feet) above the ground
- Form factor – the shape of the tree, based on recorded trees and commonly then given for calculating tree volumes for a given species. It is usually related to DBH or age class. It is distinct from taper. So it can be conical or paraboloid for example.
- Girard form class – an expression of tree taper calculated as the ratio of diameter inside the bark at 16 feet above ground to that outside bark at DBH, primary expression of tree form used in the United States
- Quadratic mean diameter – diameter of the tree that coordinates to the stand's basal area
- Site index – a species specific measure of site productivity and management options, reported as the height of dominant and co-dominant trees (site trees) in a stand at a base age such as 25, 50 and 100 years
- Tree taper – the degree to which a tree's stem or bole decreases in diameter as a function of height above ground. So it can be sharp or gradual.

Volume Estimation

- Stocking – a quantitative measure of the area occupied by trees relative to an optimum or desired level of density, which will vary according to management purpose
- Stand Density Index – a measure of the stocking of a stand of trees based on the number of trees per unit area and DBH of the tree of average basal area, based

on historical records and is particular to site type and species

- Volume table – a chart based on volume equations that uses correlations between certain aspects of a tree to estimate the standing volume
- Stand density management diagram – model that uses current stand density to project future stand composition

Volume can be calculated from the metrics recorded in a plot sample. For example, if a tree was measured to be 20m tall and with a DBH of 19 cm using previous measured tree data a volume could be approximated according to species. Such a table has been constructed by Josef Pollanschütz in Austria.

Volume of tree = BA X h x f pollanschutz

So f pollanschütz would be derived from the table and is properly called the Form Factor.

To scale this up to a hectare level the result would have to be multiplied by the number of trees of that size. This is called the blow up factor.

Tools used in Inventory

Biltmore stick – utilizes ocular trigonometry to quickly measure diameter and height

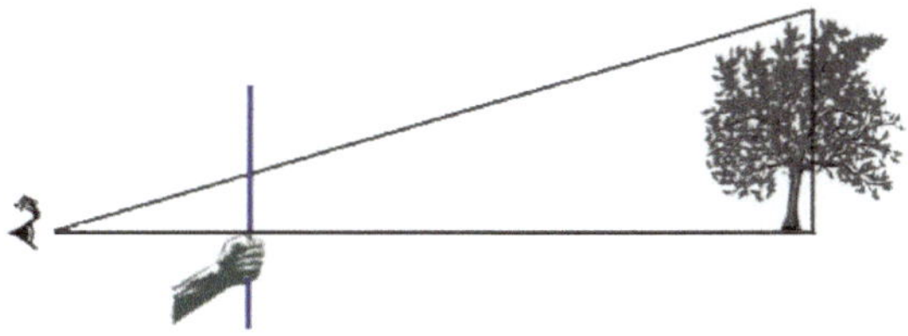

Figure demonstrating the ocular trigonometric principles behind the Biltmore stick.

- Diameter tape – cloth or metal tape that is wrapped around the bole, scaled to diameter
- Caliper – two prongs connected to a measuring tape are placed around the most average part of the bole to determine diameter
- Relascope – multiple-use tool that is able to find tree height, basal area, and tree diameter anywhere along the bole
- Clinometer – common tool used to measure changes in elevation and tree height
- Cruising rod – similar to a caliper, calculates the number of pieces of lumber yielded by a given piece of timber by measuring its diameter
- LASER Scanner used with computer software to calculate the metrics from the collected data by use of Lidar.

- Wedge prism – a small glass wedge that refracts light to create visible offsets in order to be able to choose which trees at a sampling point should be included in the sample.
- Data collector – an electronic device used to quickly enter sample data, geo-locate the data, and, in more modern times, to also access reference, web and historic materials while timber cruising.
- Increment borer – a device used to retrieve a cylindrical sample of wood material orthogonally from the stem while doing as little damage as possible to the remaining tissues.

In 2014, the Food and Agriculture Organization of the United Nations and partners, with the support of the Government of Finland, launched Open Foris – a set of open-source software tools that assist countries in gathering, producing and disseminating reliable information on the state of forest resources. The tools support the entire inventory lifecycle, from needs assessment, design, planning, field data collection and management, estimation analysis, and dissemination. Remote sensing image processing tools are included, as well as tools for international reporting for REDD+ MRV and FAO's Global Forest Resource Assessments.

References

- Sustainable Financing for Forest and Landscape Restoration - Opportunities, challenges and the way forward (PDF). Food and Agriculture Organization of the United Nations and the Global Mechanism of the UNCCD. June 2015. ISBN 978-92-5-108992-7.
- Nathan, Susan (2005). The Other Side of Israel: My Journey Across the Jewish/Arab Divide. New York: Nan A. Talese. pp. 130–131. ISBN 978-0-385-51456-9.
- Nations, Food and Agriculture Organization of the United. "Assisted natural regeneration of forests". www.fao.org. Retrieved 2016-02-24.
- "City of Ashland, Oregon - Ashland Forest Resiliency Project - What is AFR?". www.ashland.or.us. Retrieved 2016-02-05.
- "Ashland Forest Resiliency Project - Lomakatsi Restoration Project". Lomakatsi Restoration Project. Retrieved 2016-02-05.
- Gillis, Justin (2016-05-16). "In Latin America, Forests May Rise to Challenge of Carbon Dioxide". NYT. Retrieved 2016-05-17.
- "In Latin America, Forests May Rise to Challenge of Carbon Dioxide". New York Times. 16 May 2016. Retrieved 18 July 2016.
- "Philippines 5th out of 234 countries on forest gain". The Philippine Star. February 22, 2016. Retrieved February 22, 2016.
- "India should follow China to find a way out of the woods on saving forest people". The Guardian. 22 July 2016. Retrieved 7 August 2016.

7

Global Initiatives for Forest Conservation

The initiative taken for forest conservation is world forestry congress, billion-tree campaign and international day of forests. One of the largest assemblies of the world's forestry sector is the world forestry congress. The billion-tree campaign was introduced because of the alarming increase in global warming and climate change. The target of this challenge has been to plant one billion trees. This chapter discusses topics that are of great importance to broaden the existing knowledge on forest conservation.

World Forestry Congress

The World Forestry Congress (WFC) is the largest and most significant gathering of the world's forestry sector and it has been held every six years since 1926 under the auspices of the Food and Agriculture Organization of the United Nations (FAO), organized by the government of the host country. It is a forum for the sharing of knowledge and experience regarding the conservation, management and use of the world's forests, and covers such issues as international dialogue, socio-economic and institutional aspects, and forest policies.

The aim of the WFC is to contribute to orient/shape international action in silvicultural aspects and stimulate global competency and reflection that can inform the development of the global forestry sector. The WFC may also be providing guidance to national and international forestry organizations as to the most appropriate and effective means to coordinate international actions concerning technical and policy. Finally, the Congress can contribute to identify action lines, define guidelines to stimulate the development of national policies, propose incentives to orient the public and private sector, promote international cooperation in forestry matters, whilst contributing to social progress and the conservation of natural resources at global, regional and national levels.

General Relevance and Objectives

The World Forestry Congresses are an international effort aimed at influencing and stimulating reflection and analysis of the different factors, conditions, techniques, methodologies as well as professional or personal positions, thinking and ideas concerning global silviculture science. As such, each WFC has contributed to create awareness, to review and/or to formulate new approaches to technical, scientific or policy actions within the forestry sector.

The value and utility of the WFC rests also in the opportunity it offers to the world foresters and sectoral decisions-makers to analyse and expose their technical convictions and experiences, to discuss and exchange ideas and thinking, and to broaden their perspectives by sharing their common as well as diverse problems. This is particularly true today, as forestry development is becoming more and more a topic with international linkages and implications.

The growth in exchanges of knowledge and people in the forestry sector benefits all countries, and especially those with more need for forestry development. The diffusion of the scientific and technical advances as well as the human contacts promoted by the Congress, contribute to a better global balance in this sector and allow dissemination of more accurate information about forestry.

The WFC does not play the role of assessing the political, economic and social objectives of the participant countries. However, in its capacity as extraordinary assembly with broad representation from many countries and forestry groups, it contributes continuously to drawing public attention to forest related issues, to the need to raise livelihoods, and to the importance of the multiple benefits provided by forests.

Future WFCs will continue to promote international exchange, both as a result of new research and collaborations and of better comprehension of forest ecosystems and techniques for the sustainable development of these resources in the 21st century.

List of World Forestry Congresses

1st and 2nd World Forestry Congresses: 1926 and 1936

The 1st World Forestry Congress was held in Rome in 1926.

The 2nd World Forestry Congress was convened at Budapest in 1936. These two conferences have been milestones in the development of international co-operation in forestry. The impetus which they gave has become evident in the ever-growing common effort for a solution of the many problems affecting forestry and forest products. The Third Congress was planned for 1940 with the Government of Finland as host. The war, however, intervened and the Congress had to be postponed.

3rd World Forestry Congress: Helsinki, Finland, 1949

At the end of World War II, the Government of Finland was again prepared to extend its hospitality to a gathering of the world's foresters. The 1947 session of FAO's annual Conference, held at Geneva, passed a resolution calling for a World Forestry Congress in 1949. FAO's Annual Conference gladly accepted Finland's proposal of Helsinki resuming its role as the site of the Third World Forestry Congress. Professor E. Saari was appointed Chairman of the Organizing Committee and Mr Leppo, Secretary-General.

As a memorial to the Third World Forestry Congress, a curly birch was ceremoniously planted in the courtyard of the Forestry Building in Helsinki. In front of the tree stands the plaque reproduced on the cover. The planting was performed by Professor E. Saari, President of the Congress; Mr Charles Granger, United States; and Mr T. Petrov, USSR

"When this third Congress reviews the progress made since 1936, it will be seen how greatly ideas and practice in various countries have benefited from international exchanges of view and experience."

The Congress achieved an important resolution, recommended the preparation of a forestry dictionary in the major languages of the world. Such a dictionary should not only list translations of terms; it should also give precise definitions of the terms most commonly used in forestry and of the more important phrases of the forester's vocabulary.

The main themes of the Third Congress were:

- Silvics and silviculture
- Forest surveys (the essential role of the techniques of sampling and forest inventories in the composite picture of forestry and wood technology)
- Forest economics (including Forest policy)
- Forest utilization
- Forest industries

4th World Forestry Congress: Dehra Dun, India, 1954

The objective of the Congress was to determine the role and place of forested areas in the general land economy and economic development of a country, in the light of progress made to date in knowledge of forest resources, in silviculture and management, in logging and utilization.

The main themes for discussion were:

1. Present situation of forest protection and forest management in the world
2. Protective functions of the forest
3. Forest products utilization
4. Tropical forestry

Read the following excerpt from FAO's Director-General Message to the Fourth World Forestry Congress:

"What do I expect from the Fourth World Forestry Congress? Not directives which only Governments can give, but instructive information and advice on which those directives can be based and, following which, I can see that our abilities for service can be made most effective."

If we can recognize the inviting horizons ahead and be advised how we can move towards them in helpful association with each other, I am convinced we can realize our ambitions. In so doing we can experience the gratification of achievement which, after all, is perhaps one of the highest forms of compensation to be derived from any endeavor."

Cover photograph. The special commemorative stamp honoring the **Fifth World Forestry Congress**, issued by the Postal Administration of the United Nations. First day issues on sale in New York City and Seattle on 29 August. Foresters everywhere are grateful to the Secretary- General of the United Nations, Mr **Dag Hammarskjöld**, for this signal tribute to forestry.

5th World Forestry Congress: Seattle, United States of America, 1960

The central theme of the Congress was the "Multiple Use of Forest and Associated Lands." Multiple use means the management of the forest in a manner that, while conserving the basic land resource, will yield a high level of production in the five major uses - wood, water, forage, recreation, and wildlife - for the long-run benefit of the largest possible number of people.

The main themes discussed in the Congress were:

1. Forest Protection
2. Education
3. Forest and Range Watersheds
4. Logging and Forest Operations
5. Genetics and Tree Improvement
6. Forest Economics and Policy
7. Forest Products
8. Forest Recreations and Wildlife
9. Tropical Forestry

The Congress concluded that the multiple-use concept presented new challenges and new opportunities to foresters everywhere, with great prospects for additional services to the welfare of mankind. It was not, however, a panacea which could solve all forest management problems. Often one use must be dominant and other proposed uses must not then be detrimental to the major one. Moreover, there can be disadvantages to multiple use when it is inefficiently applied. But multiple use must be recognized as an important goal of forest policies.

The Congress was attended by 1,970 participants representing 65 countries.

6th World Forestry Congress: Madrid, Spain, 1966

The main theme of the Congress was "The role of forestry in the changing world economy", based on a study conducted for FAO by Wood: "World trends and prospects."

In November 1965, representatives from 114 Member States met at FAO's headquarters in Rome, under the chairmanship of Maurice Sauvé, Minister of Forestry, Canada, for the thirteenth session of the Conference, and celebrated the Twentieth Anniversary of the founding of FAO.

Reviewing the world forestry situation, the Conference set down certain considerations which must determine the scale and pattern of FAO's future activities. Among these was the following:

"The forestry administrator today faces tasks considerably more complex than those

faced by his predecessors. This is particularly true in developing countries. The number of institutions for higher forestry education responding to these changing needs is as yet insufficient. FAO is uniquely placed to interpret these changing needs, through its intimate contacts with national forest administrations, its working relations with forest industries, its awareness of the requirements of developing countries, and its contacts with, and responsibility for, many forestry educational institutions. On FAO therefore, lies the responsibility for promoting discussion and initiating action, by assisting forestry schools to transform their graduate and postgraduate training."

The main themes discussed in the Congress were:

1. World trends in wood resources and requirements
2. Planning the use of forest potentials
3. Institutional framework for forestry development
4. Financing forestry and forest industries development

The Congress urged governments receiving aid from outside sources to create by their own efforts and endeavours such conditions as would enable them best to profit from the technical and material assistance afforded to them. The Congress hoped these kinds of assistance could then be further expanded.

The Congress was attended by 2,700 participants representing 93 countries.

Cover. The 1972 FAO Ceres medal is by the French medallist Georges Simon, from a design by Mario Cassola of FAO on the analogy "Forestry is for people." The obverse side represents Indira Gandhi of India, as a stateswoman having some of the heaviest responsibilities of modern times. Ceres was the Roman Goddess of Agriculture.

7th World Forestry Congress: Buenos Aires, Argentina, 1972

The main theme of the Congress was: "Forests and socioeconomic development."

The congress examined its multi-facet challenges: how to accelerate economic and social progress whilst maintaining or enhancing the quality of the environment, recognizing that the aspiration to raise living standards often finds expression in forms that threaten the environment:

The Congress believes that the Plan of Action formulated by the United Nations Conference on the Human Environment at Stockholm in 1972 will influence forestry development throughout the world in the years to come. Recognizing that in many countries declared forest policies are not in accord with new knowledge, new preoccupations and new aspirations, the Congress considers it is now urgent to redefine forest policies in view of these new circumstances. The Congress firmly believes that, whatever the political objectives, whatever the form of economic organization, whatever the present pattern of forest land tenure, governments are responsible for planning the continuous flow of the productive, protective and social goods and services from the forest, ensuring that the physical output and environmental benefits of the forests are available for the general welfare of their peoples now and for all time. Since we live in one world, and since the forest resources of the world are unevenly distributed, national policies and plans must take account of the international context.

The central phase of the congress was devoted to technical discussions extending over the whole range of the professional interests of the participants. The main themes were:

1. Silviculturists
2. Professors, teachers and students
3. Conservationists and recreationists
4. Loggers
5. Researchers
6. Economists, administrators and planners
7. Industrialists

The congress received representation from over 80 countries. This was in part made possible by the granting of fellowships by several national technical aid agencies and by the host country.

Named as co-presidents were the heads of forest administrations in Argentina's neighbouring countries where study tours were conducted: Brazil, Chile and Uruguay.

8th World Forestry Congress: Jakarta, Indonesia, 1978

The Congress, with the theme "Forests for People", examined in depth how forestry might best serve human beings, individually and collectively. As a consequence, the

Congress declared that the world's forests must be maintained, on a sustainable basis, for the use and enjoyment of all people.

The main themes analyzed by the Congress lead to:

1. Reaffirming the responsibility of governments and on their forest administrations to manage every hectare of forest to best advantage, at the interest of all world, to avoid the forest disappearance specially of the tropical forests
2. Confirming that and because of the close and direct link between quality and quantity of water on the one hand and the extent and state of the forest resource on the other, every effort should be made in order to prevent water becoming so scarce as to affect biological production adversely
3. Recognizing the gravity of world food shortages, for which an enlarged concept of multiple-use forestry is required, which encompasses the more direct production of food from forests and forest trees, as well as from wild fauna
4. Reaffirming that the world watershed management problems require major attention and awareness in view of the existent technological means to solve them
5. Recognizing that the energy crisis and the increased oil prices will have a further and potential long term implication for forestry, with the possibility that forests become an important source of energy.

Over 2,000 participants from 100 countries and 17 international organizations attended the Congress.

The Congress strongly emphasized the important role which forest products other than timber play in the life of rural communities depending on the forests. These products play a vital role in providing employment opportunities and raising the living standard of rural people, increasing the number of people supported by forests. The Congress therefore recommended that forest services and international organizations should work for the conservation, promotion and rational utilization of such products.

9th World Forestry Congress: Mexico City, Mexico, 1985

The year 1985 had been declared "International Year of the Forest" by the FAO Council. Year 1985 also marked FAO's fortieth anniversary, as well as the occasion of the Ninth World Forestry Congress in Mexico City. The theme of the Congress was "Forest Resources in the Integral Development of Society."

It emphasized the need for foresters to look beyond their own resource base to consider how forest policies can be formulated in the light of such global events as an international banking crisis. The purpose of the Congress was to develop clear lines of action that may serve as guidelines for (a) formulating national policies; (b) providing incentives and guidance to the private and government sectors; and (c) promoting international cooperation in forestry.

During the 20 sessions of the three technical commissions into which the Congress was divided, 20 agenda items were discussed, with the presentation of 22 basic documents and 160 special documents. The long list of subjects studied included:

1. The increasing destruction of tropical forests
2. The dramatic shortage of fuelwood in arid and semi-arid zones
3. The threat posed by degraded mountain areas
4. The atmospheric pollution of forests, and the deterioration of the environment in highly industrialized countries
5. The forest fires
6. The new advances in the fields of remote sensing, forest genetics, wood as a source of energy, prevention and control of forest fires, forest hydrology and industrial uses of timber.

Besides the sessions of the Congress itself, a series of satellite meetings was held, including:

- The Expert Consultation on the Role of Forestry in Combating Desertification
- The International Forum of Forestry Youth
- The meeting on Prospects for International Financing for Forestry Programmes
- The first International Symposium on Forest Fauna
- The thirteenth session of the FAO Advisory Committee on Forestry Education
- The Latin American Round Table on Forest Genetic Resources
- The Meeting of the International Society of Tropical Foresters.

Over 2,000 participants from 105 countries attended the Congress, including representatives of many national and international organizations, institutions and associations and a large number of private individuals. Amongst the participants were professional foresters, scientists, educators, students, politicians, industry officers, workers and bankers.

10th World Forestry Congress: Paris, France, 1991

The overall theme of the Congress was "Forests, a heritage for the future". Congress members confronted the difficult task of reconciling conservation of the resource base with the need for its rational use in sustainable development.

The main themes of the Congress were as follows:

1. The forest, a protective heritage
2. Conservation and protection of the forest heritage
3. Trees and forests in rural and urban land management
4. Management of the forest heritage
5. The forest heritage, an economic resource
6. Policy and institutions.

Read the following excerpt form an interview with FAO's Director-General, Edouard Saouma:

"The threat to forests is worldwide and by no means confined to tropical forests. In short, we run the very real risk of undermining a resource base that is fundamental to the future development of the Earth. This must not be allowed to occur. Our responsibility extends not only to all people now living on the planet, but also to future generations. Therefore, I cannot but applaud the selection of "Forests, a heritage for the future" as the theme for the Tenth World Forestry Congress. As co-organizer of the Congress, along with the Government of France, FAO has committed substantial human and financial resources to ensuring that this will be the most widely followed Congress to date. The fruits of this effort are already evident; the number of voluntary papers was forecast as approximately 300 but more than 700 have been received. In the same vein, I would also wish to note that FAO has chosen "Trees for life" as the theme for World Food Day, 16 October 1991."

More than 2,500 forestry decision-makers from 136 countries attended the Congress.

11th World Forestry Congress: Antalya, Turkey, 1997

The general theme of the Congress was "Forestry for Sustainable Development: Towards the Twenty-first Century". The theme amply reflects the unique opportunity offered to the Congress and to the forestry sector at large to provide technical responses to the political issues raised in these fora.

In fact, as the last of these gatherings last century, the Eleventh WFC carried with it the special responsibility of summing up what had been achieved and of looking ahead to the challenges of the third millennium. Against this background, the general theme was selected in clear recognition of the importance of viewing forestry not as an isolated technical discipline, but rather as an important component of overall socio-economic development.

The technical sessions of the Congress were organized around the following main programme areas:

1. Forest and Tree Resources

2. Forests, Biological Diversity and Maintenance of Natural Heritage
3. Protective and Environmental Functions of Forests
4. Productive Functions of Forests
5. Economic Contribution of Forestry to Sustainable Development
6. Social Dimensions of Forestry's Contribution to Sustainable Development
7. Policies, Institutions and Means for Sustainable Forestry Development
8. Ecoregional Review.

The programme of the Congress presented 38 themes for discussion, each one based in a special document with a total of 1.200 voluntary papers. During the nine-day conference, delegates attended eight plenary sessions and 43 technical sessions. In addition to these sessions, three pre-Congress satellite meetings took place, an informal ministerial meeting and approximately 30 side meetings and other special events.

More than 4,400 participants from 145 countries attended the Congress, representing the scientific community, governments, international organizations, non-governmental organizations (NGOs) and the private sector.

The Antalya Declaration expressed concern about the discrepancy between the impressive progress in the development of concepts and tools, national programmes, global and regional treaties and varied partnerships among governments and international organizations, and the continued alarming rate of deforestation and degradation of forest resources, calling for a harmonization of the planet and its inhabitants' needs with the forests' potential to produce a vital support through political, institutional and government forestry strategies, also as a contribution to the UN Millennium Development Goals and other international agreements.

The Congress requested FAO to present an assessment of progress on the strategies outlined in the Final Statement to the XIII World Forestry Congress.

12th World Forestry Congress: Quebec City, Canada, 2003

The main theme of the Congress was "Forests: source of life."

The thematic areas were developed around the followings aspects:

1. Forests for people
2. Forests for the planet
3. People and forests in harmony

A total of 1,038 voluntary papers and 456 posters were received from all over the world - of which nearly 200 were selected for presentation, based on peer review by specialists in FAO and in Canadian institutions. In addition, 32 invited papers were presented, categorized as position papers and special papers. The technical programme was structured around ten plenary sessions, 38 theme sessions, poster sessions and 115 side events. Innovation with respect to preceding congresses included ten eco-regional sessions followed by round table discussions, and six open fora.

The Congress attracted 4,061 participants from more than 140 countries. The participants, representing a cross-section of society concerned with forests, included individuals from rural communities, private forest owners, labour, indigenous peoples, youth, industry, environmental and other non-governmental organizations, the scientific and academic community, various levels of government and international organizations.

The main output of the congress, the Final Statement, identifies areas of priority concern and is intended to encourage decisions and action by those involved with various aspects of forests and forestry, and in other related sectors.

The Final Statement calls for the development and dissemination of methodologies for assessing, reporting and managing the complete array of forest products; calls for action to manage forests and intensify forest landscape restoration and rehabilitation activities to support livelihoods, increase forest cover, enhance biological diversity and functionality, and minimize the impact of invasive alien species; and encourage collaborative partnerships involving women, forest owners, indigenous peoples, non-governmental organizations, local communities, industry and public agencies.

13th World Forestry Congress: Buenos Aires, Argentina, 2009

The XIII World Forestry Congress was held in Buenos Aires, Argentina from 18 to 23 October 2009.

Click **here** to access the **XIII WFC** web site

Under the theme “Forests in development: a vital balance” over 7,000 experts from governments, academia, research, the private sector, and non-governmental organizations came from all continents to discuss and exchange experiences over a seven-day period on a broad range of forestry related topics.

The Congress included presentations, discussions, round tables, parallel events and exhibitions, focusing on the following main issues:

1. Forests and biodiversity
2. Production for development
3. Forests at the service of people
4. Caring for our forests
5. Development opportunities
6. Organizing forest development
7. People and forests in harmony

The Congress addressed social, ecological and economic aspects of sustainable forest management in local, regional and global contexts.

14th World Forestry Congress: Durban, South Africa, 2015

The XIV World Forestry Congress was held in South Africa, 7–11 September 2015. The central theme of WFC XIV was *Forests and People: Investing in a Sustainable Future.* WFC XIV emphasized the role of forests in income generation, employment creation and equity and the vital links between forests and climate change, biodiversity conservation and sustainable water resources. The six Congress sub-themes included:

- forests for socio-economic development and food security;
- building resilience with forests;
- integrating forests and other land uses;
- encouraging product innovation and sustainable trade;
- monitoring forests for better decision-making and improving governance by building capacity.

Nearly 4,000 participants from 138 countries attended the Congress. The meeting produced three outcome documents in 6 languages:

- The Durban Declaration: 2050 vision for forests and forestry
- Message on Climate Change from the XIV World Forestry Congress
- Message from XIV World Forestry Congress to the United Nations General Assembly Summit for the adoption of the 2030 Agenda for Sustainable Development

Over 3,500 people from 120 countries contributed 825 technical papers and 269 posters. Thirty percent of the papers were from Africa and 30 percent written by women as lead authors.

The summary brief *Foresters call for action: Future land management needs better integration of sectors* listed ten key recommendations and action points from the Congress.

The Food and Agriculture Organization of the United Nations released key findings of an assessment of the world's forest resources in the report *Global Forest Resources Assessment 2015*.

During the Congress, Ugandan community leader Gertrude Kabusimbi Kenyangi was awarded the Collaborative Partnership on Forests 2015 Wangari Maathai 'Forest Champions' Award in recognition of her efforts to promote the conservation and sustainable use of her country's forest resources. Given in recognition of outstanding efforts to improve and sustain forests, the award was established in 2012 by the Collaborative Partnership on Forests (CPF), which is chaired by FAO. Its name honours the memory of Wangari Maathai, the Kenyan environmentalist and the first African woman to win a Nobel Peace Prize for her contribution to sustainable development, democracy and peace.

Billion Tree Campaign

The Billion Tree Campaign was launched in 2006, by the United Nations Environment Programme (UNEP) as a response to the challenges of global warming, as well as to a wider array of sustainability challenges, from water supply to biodiversity loss. Its initial target was the planting of one billion trees in 2007 (achieved by November). One year later, in 2008, the campaign's objective was raised to 7 billion trees – a target to be met by the climate change conference that was held in Copenhagen, Denmark in December 2009. Three months before the conference, the 7 billion planted trees mark had been surpassed. In December 2011, after more than 12 billion trees had been planted, UNEP formally handed management of the program over to the youth-led not-for-profit Plant-for-the-Planet Foundation, based in Tutzing, Germany. Over 14.2 billion trees have been planted as of 2016. The Billion Tree Campaign calls upon all individuals, countries and companies to participate, and accepts all contributions. Planted trees and pledges can be registered here.

Inspiration

The Billion Tree Campaign was inspired by Nobel Peace Prize Laureate Wangari Maathai, founder of the Green Belt Movement. When an executive in the United States told Professor Maathai their corporation was planning to plant a million trees, her response was: "That's great, but what we really need is to plant a billion trees." The campaign was carried out under the patronage of Prince Albert II of Monaco.

"It's the little things citizens do. That's what will make the difference. My little thing is planting trees." – Wangari Maathai

Response

Under UNEP's leadership and through proactive advocacy by the patrons and the partners, the Billion Tree Campaign catalysed tree planting action on all continents. The billionth tree, an African Olive, was planted in Ethiopia in November 2007. The two billionth tree took root as part of the United Nation's World Food Programme agroforestry initiative. The campaign's target was then raised to seven billion trees. In 2009, UNEP mobilized action across the globe through the Twitter for Trees campaign on www.twitter.com/UNEPandYou. The initiative was driven by a simple yet powerful idea: UNEP pledged to plant one tree to feed into the Billion Tree Campaign, for every follower who joined from 5 May 2009 to World Environment Day on 5 June 2009. The campaign was a success, with 10,300 people following the UNEPandYou page by World Environment Day weekend.

The World Organization of the Scouts Movement also planted trees under the campaign in line with its mandate to study and protect Nature across several countries.

The United Nations Peacekeeping missions also joined the campaign and planted trees with its field missions in East-Timor, Côte d'Ivoire, Darfur, Lebanon, Haiti, Congo, and Liberia amongst others missions.

Roll of Honour Top 10 Countries

- China 2.8 billion
- Pakistan 1 billion
- India 2.1 billion
- Ethiopia 1.6 billion
- Mexico 785 million
- Turkey 716 million
- Nigeria 612 million
- Kenya 455 million
- Peru 246 million
- Myanmar 191 million
- Cuba 137 million

"Around a flowering tree, one finds many insects." – African saying

December 2011, five years since the campaign's launch, the campaign's website www.unep.org/billiontreecampaign proudly registered over 12.5 billion planted trees across 193 countries.

Handover to the Young Generation

The Billion Tree Campaign was handed over to the Plant-for-the-Planet Foundation in December 2011, an organisation that has been participating in the Billion Tree Campaign since 2007. The momentum has continued with 40,000 young ambassadors spreading the message in over 100 countries.

Quote from UNEP

"Looking back over the Billion Tree Campaign's greatest successes, what is most remarkable is not its scale, but its spread. People from all around the world have enthusiastically joined the campaign and planted trees in their own communities." – Achim Steiner, UNEP Executive Director.

International Day of Forests

The International Day of Forests was established on the 21st day of March, by resolution of the United Nations General Assembly on November 28, 2012. Each year, various events celebrate and raise awareness of the importance of all types of forests, and trees outside forests, for the benefit of current and future generations. Countries are encouraged to undertake efforts to organize local, national, and international activities involving forests and trees, such as tree planting campaigns, on International Day of Forests. The Secretariat of the United Nations Forum on Forests, in collaboration with the Food and Agriculture Organization, facilitates the implementation of such events in collaboration with governments, the Collaborative Partnership on Forests, and international, regional and subregional organizations. International Day of Forests was observed for the first time on March 21, 2013.

Background

Each year more than 13 million hectares (32 million acres) of forests are lost, an area roughly the size of England. As go the forests, so go the plant and animal species that they embrace – 80% of all terrestrial biodiversity. Most importantly, forests play a critical role in climate change including global warming: deforestation results in 12-18 percent of the world's carbon emissions – almost equal to all the CO_2 from the global transport sector. Equally crucial, healthy forests are one of the world's primary 'carbon sinks.'

Today, forests cover more than 30% of the world's land and contain more than 60,000 tree species, many as of yet unidentified. Forests provide food, fiber, water and medicines for approximately 1.6 billion of the world's poorest people, including indigenous peoples with unique cultures.

History

In November 1971, the "States members" at the 16th session of the Conference of the Food and Agriculture Organization, voted to establish "World Forestry Day" on March 21 of each year. From 2007-2012, the Center for International Forestry Research (CIFOR) convened a series of six Forest Days, in conjunction with annual meetings of the United Nations Framework Convention on Climate Change Conference of Parties. CIFOR organized these events on behalf of and in close cooperation with other members of the Collaborative Partnership on Forests (CPF). Following the International Year of Forests in 2011, the International Day of Forests was established by resolution of the United Nations General Assembly on November 28, 2012.

Forest Day

The catalyst for Forest Day was a casual conversation in Oxford, England, in February 2007, between two scientists who felt the world was underestimating the importance of forests in mitigating carbon emissions and saw a glaring need for the latest forestry research and thinking to inform global policy makers and UNFCCC negotiators. They did not foresee that the conference would become one of the most influential global events on forests and climate change today.

Forest Day 1: Bali, Indonesia (2007)

Journalists interview Center for International Forestry Research scientist Daniel Murdiyarso

The inaugural Forest Day was one of the major events at United Nations Framework Convention on Climate Change (UNFCCC) Conference of the Parties (COP) 13 in Bali, Indonesia on 7 December 2007. More than 800 people participated in Forest Day, including scientists, members of national delegations, and representatives from intergovernmental and non-governmental organizations.

A major feature of Forest Day was four parallel panel discussions focusing on cross-cutting themes related to forests and climate change. These well-attended discussions examined such issues as setting baselines and methodological challenges in estimating forest carbon; markets and governance challenges associated with Reducing Emissions from Deforestation and Forest Degradation (REDD+); adaptation to climate change; and equity-efficiency trade-offs.

Areas of consensus emerging from the discussions included the following:

- While there are significant methodological challenges to be surmounted, current methods are "good enough" to proceed with the design of mechanisms for reducing emissions from both deforestation and degradation.
- Governance-related challenges pose the greatest risks for both international investors and local stakeholders in the context of new mechanisms.
- Mechanisms should be simple, and should not repeat the mistakes of the Clean Development Mechanism.
- The success of any REDD+ mechanism will depend on the political will to address the drivers of deforestation, including drivers that originate beyond the forestry sector.
- Adaptation efforts need to shift from responsive to anticipatory, and should be focused on the most vulnerable, including forest-dependent people.

Forest Day 2: Poznań, Poland (2008)

Sub-plenary 1, organized by IUFRO, CIFOR, ICRAF and PROFOR-World Bank, discussed the cross-cutting theme "adaptation of forests to climate change." Panelists included (L-R): Bruno Locatelli, CIFOR; Gerhard Dieterle, World Bank; Markku Kanninen, CIFOR; Balgis Osman-Elasha, IUFRO; and Dennis Garrity, World Agroforestry Center (ICRAF).

Building on the positive response to the first Forest Day, Forest Day 2, held in Poznań, Poland, on December 8, 2008, brought together nearly 900 participants to discuss the opportunities and challenges of bringing forests into global and national strategies for climate change adaptation and mitigation.

Participants stressed the broad importance of forest ecosystems and noted that forests comprise an inter-dependent web of animals, plants and microorganisms, which together provide a wide range of goods and services beyond carbon sequestration. These include biodiversity conservation, rainfall generation and products that are crucial to the livelihoods of local forest dependent and indigenous peoples as well as to the economies of many countries.

Those attending recognized the importance of building on the vast knowledge and experience that exists on sustainable forest management (SFM) and called on negotiators to consult with forest stakeholders as they develop climate policy.

Frances Seymour, Director General of CIFOR delivered a summary of Forest Day 2 to Yvo de Boer, Executive Secretary of the UNFCCC. This summary was drafted by a committee representing members of the CPF, and included points of consensus as well as points of disagreement that emerged during the course of the day. Seymour highlighted the need to:

- Include forests in climate mitigation and adaptation mechanisms and strategies
- Ensure full inclusion and participation of civil society in international, regional, national and local decision-making processes
- Recognize and respect the rights of women, poor people and Indigenous Peoples

Forest Day 3: Copenhagen, Denmark (2009)

Nearly 1500 stakeholders attended Forest Day 3 held in Copenhagen, Denmark on 13 December 2009 including 34 donors, government representatives, 88 journalists, 500 NGO representatives, indigenous leaders, 188 private sector representatives and hundreds of scientists and forestry experts. Their goal was to ensure that the design and implementation of forest-related climate mitigation and adaptation measures under consideration in the Climate Change Agreement would be effective, efficient and equitable.

Although the UNFCCC failed to agree on binding targets for reducing greenhouse gas emissions, significant progress was achieved in negotiating the outlines of a REDD+ mechanism. The Copenhagen Accord that emerged was the first international agreement to recommend that financial resources be raised to support REDD+. Australia, France, Japan, Norway, the United Kingdom and the United States offered a US$3.5 billion funding package for REDD+ preparation.

Moreover, one of the indicators of the relevance of Forest Day – its ability to attract world leaders – became clear at Copenhagen. Key speakers at the event included:

- Rajendra K. Pachauri, Chair of the Intergovernmental Panel on Climate Change
- Wangari Maathai, founder of the Green Belt Movement and Nobel laureate

- Gro Harlem Brundtland, former Norwegian Prime Minister
- Hilary Benn, the UK Secretary of State for Environment, Food and Rural Affairs
- Sir Nicholas Stern, head of the Grantham Research Institute on Climate Change and the Environment.

Former US President Bill Clinton appeared via video and Nobel Prize winner Elinor Ostrom was among the keynote speakers.

Forest Day 4: Cancún, Mexico (2010)

Daniel Nepstad during Forest Day 4, Cancún, Mexico

Forest Day 4 took place on 5 December 2010 in Cancún, Mexico. More than 1,500 people including over 280 Climate negotiators were in attendance.

The theme for Forest Day 4 was "Time to Act," highlighting the urgency of ensuring the survival of the world's forests, the biodiversity they embrace and the hundreds of millions of people who depend on them. The event served as a bridge between the International Year of Biodiversity and the 2011 International Year of Forests. Forest Day 4 was hosted by the Government of Mexico, through the National Forestry Commission (CONAFOR), CPF and CIFOR.

President Felipe Calderón Hinojosa, in his opening speech at FD4, told the plenary "... it's time for all of us to push, and push hard for the full incorporation of REDD+ into a long-term international climate change agreement." In a passionate plea, the Mexican president also stressed, "Either we change our way of life now, or climate change will change it for us."

Keynote speeches by Daniel Nepstad, Director of the International Program at the Amazon Environmental Research Institute and Mirna Cunningham Kain, Chair of the Center for Autonomy and Development of Indigenous Peoples, emphasized the critical needs for both sound forestry and climate change science, and pro-active engagement with indigenous people and forest-dependent communities as the de facto and de jure custodians of land and forest resources to secure equitable outcomes. UN Under Secretary-General for Economic and Social Affairs Sha Zukang provided an important forward-looking perspective, drawing participants' attention to the multiple benefits of forests and the International Year of Forests in 2011.

Key issues discussed at FD4 included:

- Harnessing REDD+ to sustainably manage forests and reduce poverty: The day saw the emergence of a robust consensus that the risks of no action on protecting the world's forests are far greater than the risks of moving ahead with less-than-perfect agreements.
- REDD+ provides a key and cost-effective opportunity to mitigate climate change: FD4 participants reaffirmed that through REDD+ we may significantly reduce, remove and avoid global emissions at reasonable cost, as long as we take due account of the rights and livelihoods of indigenous people and local communities, biodiversity and ecosystem services, whilst assisting developing forest countries adapt to climate change.
- The rights of indigenous people and forest-dependent communities need to be protected: Communities would be more willing to engage with REDD+ initiatives if they were to participate in all aspects of REDD+ design and implementation, if they are granted rights to the carbon in their forests, if they play a central role in the design of local rules, and if REDD+ does not permit more powerful competitors to threaten local interests.
- Additional financing is needed to implement REDD+ at scale: An agreement on a robust and predictable system for mobilizing financial resources from various sources is needed, primarily in developed countries. This will be needed to stimulate and pay for early REDD+ action at scale, technology transfers, capacity building and the development of national and sub-national monitoring, reporting and verification (MRV) systems, among others.
- Biodiversity conservation is a prerequisite for the success of REDD+: More than 90 percent of FD4 participants polled said that biodiversity safeguards are either "very important" or "essential" for the success of REDD+, and more than 95 percent said that it is important to monitor co-benefits.
- REDD+ and agricultural drivers of deforestation: FD4 participants proposed several options to increase agricultural intensification whilst reducing net annual rates of deforestation including: increasing production efficiencies; pro-

moting multifunctional landscapes; directing REDD+ financing to increase efficiencies in agronomic practices; and shifting extensive production systems to low carbon landscapes.

- Promoting synergies between climate change mitigation and adaptation across landscapes: More research is needed to explore linkages between adaptation and mitigation in forests at different scales. Two specific challenges – agreement on monitoring, reporting and verification (MRV) systems, and continued improvements in forest governance were also discussed.
- Strengthen linkages between national and subnational MRV systems for REDD+: Challenges remain for monitoring forest degradation and peatland emissions but these may be overcome with significant additional capacity building and technology transfer, including the application of novel technologies. Examples include Google's Earth Engine (demonstrated by Rebecca Moore, Head of Google's Global Outreach Program prior to the Closing Plenary of FD4), Open Data Kit and androids for forest biomass measurements with communities.
- Improve accounting rules for forest management in developed countries: More comprehensive accounting on forests will be beneficial both for the climate and for forests. Participants acknowledged that this is a complex and controversial issue, but success in reaching agreement on new rules would help smooth the way for agreement on new emission reduction commitments by developing countries for the second commitment period of the Kyoto Protocol.
- The success of REDD+ strategies and projects will depend on whether they influence governance reforms or are shaped by existing governance failures: The ability of developing countries to enhance the role of their forest resources in mitigating climate change is closely linked with their commitment to governance reform. There are also huge challenges to reform the embedded structures of past poor governance. However, the legality and legitimacy of REDD+ are likely to depend on a balance between central oversight and decentralized decision making, clear tenure and transparent and equitable benefit-sharing arrangements.
- Reinforcing the UNFCCC momentum on forestry and climate change : A key challenge of COP16 in Cancún is how existing mitigation and financing pledges made through the Copenhagen Accord can now be transformed into official commitments under the UNFCCC. This may require abandoning the "nothing is agreed until everything is agreed" global climate change architecture approach to ensure that a balanced cluster of decisions can be made in negotiating areas that remain close to agreement. This will ensure that commitments can be translated into actions in 2011.

Comments

- Virgilio Viana, CEO of Amazonas Sustainable Foundation, Brazil: "Forest Day has become a venue for in depth discussion relating to forests and climate change, filling a gap in the meeting spaces. A source of inspiration for policy making at national and international level."
- Yemi Katerere, UN REDD: "FD provides the forum, agenda and a theme to draw together practitioners, policy makers, researchers and NGOs involved in forestry. FD is the place to get the experts' view on forestry at COP... The CPF has a broad range of constituents; FD reflects this multi-faceted range of stakeholders."
- Felician Kilahama, Ministry of Natural Resources and Tourism, Tanzania: "FD is unique: It brings environmental stakeholders, policy makers, private sector, etc together. For some countries, FD provides a rare chance to interact internationally, exchange knowledge, lessons learned on forests and climate change."
- David Diaz, Ecosystem Marketplace (Forest Trends), USA: "There's a lot of positive energy surrounding forestry right now and FD channels that energy. President Calderon's presence has an impact. The participation of high profile figures means that other presidential/ministerial level people may get a louder voice or take an interest at that level."
- Tony la Vina, Dean, Ateneo School of Government, Manila University, Philippines: "FD is important because of the range of people who come: negotiators, practitioners, policy makers, advocates – the range is impressive. Here you can help shape the agenda for forestry. I would like to see a strong appeal to negotiators that things are already running on the ground. There's a sense of urgency. The negotiations need to be relevant to things already happening on the ground."
- Niels Elers Koch, President, International Union of Forest Research Organizations: "Forest Day (FD) is the single most important science–policy interface in the forest sector, where researchers and policy makers meet."

Forest Day 5: Durban, South Africa (2011)

The 2011 COP 17 was hosted by Durban, South Africa, from 28 November to 9 December 2011. Forest Day 5 took place on 4 December 2011 in Durban, South Africa. CIFOR convened the event, which was co-hosted by 11 members of the Collaborative Partnership on Forests and the Government of South Africa through the Department of Agriculture, Forestry and Fisheries. More than 1,100 people from 82 countries attended, including 214 official climate-change negotiators and 65 media representatives.

The theme for Forest Day 5 was 'From Policy to Practice', aiming to inform the COP agenda and forest stakeholders on ways to implement the REDD+ agreement reached in Cancún in 2010 to produce social and environmental benefits, and to integrate forests into adaptation strategies on the ground. Organisers gave special attention to issues relevant to sub-Saharan Africa, to REDD+ issues facing the humid tropical forests of the Congo Basin, and to the improved management and use of Africa's dry forests areas.

The link between forests and food security was also one of the main themes of the day. The organisers of Forest Day 5 and Agriculture and Rural Development Day (which was held on 3 December 2012) coordinated in the months leading up to the events to identify opportunities for tie-ups between the two events. A number of joint, similar-themed discussion forums were held on climate-smart agriculture, agroforestry and food security.

As in past years, Forest Day 5 attracted several high-level speakers. In all, more than 60 speakers and panellists took part in Forest Day 5, with six of the ten keynotes presented by women.

Forest Day 6: Doha, Qatar (2012)

The 2012 UNFCCC COP 18 was hosted by Doha, Qatar from 26 November to 7 December 2012. Forest Day 6 took place on the sidelines of COP 18, on 2 to 3 December 2012, and considered issues ranging from REDD+ financing to adaptation, desertification, reforestation and afforestation.

International Day of Forests

2013

The inaugural International Day of Forests "was celebrated around the world through tree-planting and other community-level events, including art, photo and film as well as social media outreach."

2014

In 2014, the International Day of Forests focused on "each individual's personal and unique connection with Forests", through a campaign entitled "My Forest | Our Future". A special event was held at the United Nations headquarters on "Women as agents of change for forests and sustainable development".

2015

The 2015 theme for the International Day of Forests is "Forests | Climate | Change".

2016

The theme selected to mark 2016's International Day of Forests was forests and water.

Forests are key to the planet's supply of freshwater. Over 100 events were held in 55 countries to celebrate the Day. In Rome, FAO headquarters, a special event was held to highlight forests' crucial role in contributing to water and food security.

References

- "Foresters call for action: Future land management needs better integration of sectors. Ten recommendations from the XIV World Forestry Congress". XIV World Forestry Congress. FAO. Retrieved 4 April 2016.
- "In the last 25 years, the world lost a forested area the size of South Africa". The Washington Post. 8 September 2016. Retrieved 4 April 2016.
- "21 Reasons To Celebrate The Value Of Trees In Honor Of International Day Of Forests". Huffington Post. Retrieved 21 November 2014.
- "Unasylva - Vol. 8, No. 3 - Fourth World Forestry Congress - Program of the Fourth World Forestry Congress". Fao.org. Retrieved 2013-05-08.
- UNEP. (7 December 2011). "UNEP Billion Tree Campaign Hands Over to the Young People of the Plant-for-the-Planet Foundation." Accessed: 26 September 2013.
- Holmgren, Peter. 2013, March 21. "Sharing positive views about forests and trees on the International Day of Forests," CIFOR. Accessed: March 21, 2013.
- "Unasylva - No. 190-191 - Eleventh world forestry congress - Table of Contents". Fao.org. Retrieved 2013-05-08.

Permissions

All chapters in this book are published with permission under the Creative Commons Attribution Share Alike License or equivalent. Every chapter published in this book has been scrutinized by our experts. Their significance has been extensively debated. The topics covered herein carry significant information for a comprehensive understanding. They may even be implemented as practical applications or may be referred to as a beginning point for further studies.

We would like to thank the editorial team for lending their expertise to make the book truly unique. They have played a crucial role in the development of this book. Without their invaluable contributions this book wouldn't have been possible. They have made vital efforts to compile up to date information on the varied aspects of this subject to make this book a valuable addition to the collection of many professionals and students.

This book was conceptualized with the vision of imparting up-to-date and integrated information in this field. To ensure the same, a matchless editorial board was set up. Every individual on the board went through rigorous rounds of assessment to prove their worth. After which they invested a large part of their time researching and compiling the most relevant data for our readers.

The editorial board has been involved in producing this book since its inception. They have spent rigorous hours researching and exploring the diverse topics which have resulted in the successful publishing of this book. They have passed on their knowledge of decades through this book. To expedite this challenging task, the publisher supported the team at every step. A small team of assistant editors was also appointed to further simplify the editing procedure and attain best results for the readers.

Apart from the editorial board, the designing team has also invested a significant amount of their time in understanding the subject and creating the most relevant covers. They scrutinized every image to scout for the most suitable representation of the subject and create an appropriate cover for the book.

The publishing team has been an ardent support to the editorial, designing and production team. Their endless efforts to recruit the best for this project, has resulted in the accomplishment of this book. They are a veteran in the field of academics and their pool of knowledge is as vast as their experience in printing. Their expertise and guidance has proved useful at every step. Their uncompromising quality standards have made this book an exceptional effort. Their encouragement from time to time has been an inspiration for everyone.

The publisher and the editorial board hope that this book will prove to be a valuable piece of knowledge for students, practitioners and scholars across the globe.

Index